ENVIRONMENTALLY SAFE DRILLING PRACTICES

ENVIRONMENTALLY SAFE DRILLING PRACTICES

PENNWELL PUBLISHING COMPANY
TULSA, OKLAHOMA

PennWell Publishing Company
1421 South Sheridan/P.O. Box 1260
Tulsa, Oklahoma 74101

Printed in the United States of America

1 2 3 4 5 99 98 97 96 95

Table of Contents

List of Illustrations and Tables

Measurements

parts per million	ppm
foot, feet	ft
square feet	sq ft
acres	ac
inch, inches	in.
hour, per hour	hr,/hr
year	yr
pound	lb
gallon	gal
quart	qt
barrel	bbl
gallons per minute	gal/min
miles	mi
British thermal unit	BTU

Preface

This book is not intended to be a comprehensive overview of environmental guidelines for all drilling practices.

Much of what is contained herein reflects the author's industry experience, with an emphasis on U.S. Lower 48 onshore and wetlands drilling operations.

While that includes dealing with regulations and practices under some of the most stringent regulatory regimes — notably South Louisiana — the reader should look to supplemental sources of information for drilling under special circumstances, such as in arctic environments. Alaskan North Slope operators and drilling contractors must cope with a very different range of concerns and regulatory oversight, and that could be a book in itself.

The same could be said for offshore drilling, where dealing with the regulatory constraints of federal law implemented by the Minerals Management Service justifies a separate volume.

At the same time, it must be said that the preferred practices and recommendations outlined in this book can serve as useful guidelines for operators and drilling contractors elsewhere in the world, notably in environmentally sensitive areas of developing countries that have little historical regulatory precedent.

Remember: If it is an environmental regulation affecting the oil and gas industry in the U.S. today, it is likely to be one implemented in another country tomorrow. So it helps to change company and individual attitudes toward drilling with care for the environment, no matter how much of today's drilling dollars are being shifted overseas.

Armando Navarro
November 1994

INTRODUCTION

With today's growing awareness of social responsibility and public interest, environmental concerns are receiving the highest priority in U.S. government regulations. The future of our environment — and the industry — depends on meeting the challenge of maintaining a balance between acceptable environmental practices and economically efficient operating procedures.

These growing concerns have imposed a burden on the oil and gas drilling industry to implement programs and techniques to not only minimize the potential impact that oil and gas operations can have on the environment but also remediate those impacts, whether they occurred through accident or negligence.

The mandate to the oil and gas industry of implementing environmentally safe drilling techniques has proven to be not only a financial challenge but also an administrative challenge.

A number of industry studies have shown that the increased scrutiny by government agencies could cost the U.S. oil and gas industry hundreds of millions of dollars in added drilling costs alone by the turn of the century. With the low prices that oil and gas products bring on today's market, these increased environmental costs will render many projects uneconomic.

Regulations governing the oil industry with regard to the environment continue to change, yet many firms that recognize they no longer can conduct their operations as they did in "the good old days" have had to employ full-time environmental professionals to ensure that all operations conducted in the field are in compliance with the changing local, state, and federal regulations. This increased scrutiny also has prompted drilling contractors to think of ways to minimize the impact their operations have on the environment rather than just ways of how to remediate the impact once it occurs.

Some extremists would have the U.S. all but ban drilling altogether, while others would turn back the clock to unfettered ways. Due to the critical nature of oil and gas to U.S. economic well-being and the need for the U.S. to minimize its dependence on imported oil and gas and the real concern over damage environmental neglect can have on sensitive ecosystems, it is obvious that neither extreme is acceptable.

This book is intended to introduce techniques that will assist oil and gas professionals in minimizing the impact that drilling operations have on the environment. These techniques range from the extremely practical that have been in place since the beginning of the modern petroleum industry to the theoretical that are effective but expensive and difficult to implement.

Drilling And The Environment

Basic Concepts

In order to gain an understanding of the techniques that will be recommended to minimize environmental impacts of contemporary drilling practices, it is instructive to look to history.

World War II and the growing importance of the automobile in American society spawned one of many boom-and-bust cycles in the oil industry. Having the ability to supply Allied forces with enough fuel to fight the war gave the United States the edge over other countries. But a problem that had not existed before now appeared in the "new" oil field. It seemed that all of the easy finds of oil pools had been made, and dry holes had become commonplace. This made it difficult for the oil industry to keep up with the demand for oil by Allied forces and the United States industrial infrastructure. This shortage helped prop oil prices to more than $1/bbl. Although this helped producers, the increasing demand for oil warranted that new sources of oil had to be found. Reserves had to be replaced in order to ensure national security.

The enthusiasm by the government to replace vital oil reserves and by oil companies to get rich allowed production to go unrestricted and once again caused the price of oil to drop to only pennies per barrel.

These low prices forced many oil producers to disregard reservoir management in an attempt to salvage profits by producing their wells with no control. This inept management caused many oil producers to prematurely abandon many fields, sometimes leaving as much as 90% of the proven recoverable reserves unproduced. This waste of natural resources prompted the government to intervene once again and to implement the first of a series of proration laws. These laws were structured to assist in the conservation of the nation's oil reserves by minimizing waste and to protect the correlative rights of the land and mineral rights owners through reservoir management.

Many oil producers at the time felt that the proration laws were just a ploy by the government and the larger companies to drive them out of business. But in reality these laws were the government's attempt to monitor the activities of the oil companies to ensure that prudent techniques were being utilized in the production of oil in order to minimize the waste of this extremely important natural resource.

The interpretation of these conservation laws still is a source of heated debate among land and mineral owners and oil producers.

Many states developed their own monitoring agencies to monitor oil industry activities. One of the first and most powerful of these agencies is the Texas Railroad Commission (RRC). The RRC determines who can drill for oil in Texas and where in the state this can be done. It also monitors producing techniques to ensure that waste is minimized and that the distribution of oil production is equitable to mineral owners, producers, and the government.

Over time, the petroleum industry has grown into a multi-trillion dollar industry due to a 7,000% increase in the consumption of oil in the world during the past 50 years. That has spurred creation of whole branches of government to regulate its activities. Until the birth of the environmental movement in the early 1970s, all of the regulatory focus had been towards the management of production and very little, if any, towards some of the inevitable side effects of drilling operations, notably its environmental impacts.

The rise of these new agencies led to increased scrutiny of oil companies' efforts to drill for and produce oil and gas. This increased scrutiny not only came from the government but also from the general public as a result of unfortunate accidents such as the Exxon Valdez oil spill in Prince William

Sound, Alaska, in 1989, which greatly intensified environmental concern.

This increased public awareness has made it difficult for oil companies to drill in areas that are considered to have sensitive ecosystems. Areas such as Offshore California, Alaska's Arctic National Wildlife Refuge Coastal Plain, the Florida Keys, and the eastern Gulf of Mexico have all been targets of environmentalists to prohibit any further drilling for oil and gas. Companies wishing to drill for oil and gas in these areas often have had to spend a great deal of time and money in court to be allowed to drill in these areas.

This increased scrutiny is not all bad, however. This public awareness has prompted the oil industry to become more aware of how to alter operations to minimize the environmental effects of drilling and producing oil. The implementation of zero-discharge operations, mitigation measures, and the development of new environmentally safe products utilized in the oil industry are all a direct result of this increased scrutiny. Continued development of these systems and products reaffirms the belief that the needs of the environment and the needs of the oil industry can coexist.

Environmental Regulations

Much of what is driving today's environmental initiative is the involvement of the government in regulating operations. Several branches of government have been developed over the past several years to serve as "watchdogs" as well as to provide guidelines on how the oil and gas industry conducts its operations. Many of these agencies are branches of the Environmental Protection Agency (EPA), which was created to monitor the state of the environment.

Initially, intervention of the government into the activities of the oil industry was intended to prevent waste of natural resources. As the industry grew, and more companies began to enter into the competitive world of oil and gas E&P, so did the government's scale of intervention. In today's world of oil and gas E&P, the government is involved in every aspect of the business from permitting through plugging and abandonment of wells.

With recent concern over the protection of the environment from damage, the government has created agencies that not only monitor the exploration and production of oil and gas but ones that monitor the undesirable waste that is generated as a by product of oil and gas drilling.

Agencies such as the EPA, Department of Environmental Quality (DEQ), and Department of Natural Resources (DNR), just to name a few, are now widely known within the oil and gas industry.

Some oil industry officials contend that the existence of these agencies and regulations has added such a huge financial burden to the industry that it makes it economically infeasible to drill for oil and gas and thus much of the reserves of oil and gas left to be proven and developed will be left in the ground. The unfortunate reality of this situation is that if the industry had known 50 years ago of the effects some of its operations would have on the environment and the growing influence of environmentalism in American life, these practices would have been modified to avert any environmental damage, and thus most of these agencies and regulations probably never would have been conceived.

For the oil industry to coexist with a regulatory infrastructure that requires a strict adherence to the regulations, a basic understanding of the laws, regulations, and methods of operation must be had by not only the officers of the oil companies but for every person even remotely involved with operations. Failure to have the basic knowledge of these laws and regulations can lead not only to continued environmental damage, but since noncompliance is a federal offense, someone in violation of these laws is liable for civil and criminal penalties that can involve fines and prison terms.

In the oil industry, there are a number of key laws that affect daily oil field operations in a very basic way. The government agencies given authority to implement these laws can be found at the state and federal levels. These laws include the Resource Conservation and Recovery Act

TABLE 1-1

EPA's List of Exempt Exploration and Production Wastes

- Produced water.
- Drilling Fluids.
- Drill Cuttings.
- Rigwash.
- Drilling fluids and cuttings from offshore operations disposed of onshore.
- Well completion, treatment, and stimulation fluids.
- Basic sediment and water and other tank bottoms from storage facilities that hold product and exempt waste.
- Accumulated materials such as hydrocarbons, solids, sand, and emulsion from production separators, fluid treating vessels, and production impoundments.
- Pit sludges and contaminated bottoms from storage or disposal of exempt wastes.
- Workover wastes.
- Gas plant dehydration wastes, including glycol-based compounds, glycol filters, filter media, backwash, and molecular sieves.
- Gas plant sweetening wastes for sulfur removal, including amine, amine filters, amine filter media, backwash, precipitated amine sludge, iron sponge, and hydrogen sulfide scrubber liquid and sludge.
- Cooling tower blowdown.

(RCRA) of 1976, Subtitles C and D; the Comprehensive Environmental Response, Compensation and Liability Act (Cercla) of 1980, also known as "Superfund"; the Title III Superfund Amendments and Reauthorization Act (SARA) of 1986; the Clean Water Act (CWA) of 1977; the Safe Drinking Water Act (SWDA) of 1986; the Clean Air Act amendments of 1990; the National Environmental Policy Act (NEPA) of 1969; and the Endangered Species Act (ESA) of 1973. As can be seen from this list of laws, a wide variety of areas are identified that can be affected by the oil and gas industry.

RCRA regulates the management of hazardous waste from generation to its final disposition. There are basically two statutes that govern oil and gas operations: Subtitle C and Subtitle D. Subtitle C authorizes a comprehensive federal program to regulate hazardous wastes, whereas Subtitle D addresses the disposal of nonhazardous solid waste, including solid, liquid and gaseous materials.

For a waste to come under the scrutiny of RCRA, the characteristics of the waste must fall within the criteria set forth under EPA's criteria governing hazardous waste. This can be found in the Code of Federal Regulations 40 CFR Parts 260 and 261. Through these codes EPA has set parameters for corrosivity, reactivity, ignitability, and toxicity by which the waste can be determined as a hazardous waste. Wastes with these characteristics will be deemed hazardous, regardless of their origin.

In 1988, EPA made a regulatory determination to exempt from RCRA Subtitle C certain wastes generated from oil, gas, and geothermal drilling and production operations from the subtitle requirements. The amendments specifically exempt drilling fluids, produced water and other associated drilling and production wastes (Table 1-1).

- Spent filters, filter media, and backwash (assuming the filter itself is not hazardous and the residue in it is from an exempt waste stream).
- Packing fluids.
- Produced sand.
- Pipe scale, hydrocarbon solids, hydrates, and other deposits removed from piping and equipment prior to transportation.
- Hydrocarbon-bearing soil.
- Pigging wastes from gathering lines.
- Wastes from subsurface gas storage and retrieval, except for the listed nonexempt wastes.
- Constituents removed from produced water before it is injected or otherwise disposed of.
- Liquid hydrocarbons removed from the production stream but not from oil refining.
- Gases removed from the production stream, such as hydrogen sulfide and carbon dioxide, and volatilized hydrocarbons.
- Materials ejected from a producing well during the process known as blowdown.
- Waste crude oil from primary field operations and production.
- Light organics volatilized from exempt wastes in reserve pits or impoundments or production equipment.

EPA further listed certain exploration and production wastes as nonexempt — although not necessarily hazardous (Table 1-2).

TABLE 1-2

EPA's List of Exempt Exploration and Production Wastes

- Unused fracturing fluids or acids.
- Gas plant cooling tower cleaning wastes.
- Painting wastes.
- Oil and gas service company wastes, such as empty drums, drum rinsate, vacuum truck rinsate, sandblast media, painting wastes, spent solvents, spilled chemicals, and waste acids.
- Vacuum truck and drum rinsate from trucks and drums transporting or containing nonexempt waste.
- Refinery wastes.
- Liquid and solid wastes generated by crude oil and tank bottom reclaimers.
- Used equipment lubrication oils.
- Waste compressor oil, filters, and blowdown.
- Used hydraulic fluids.
- Waste solvents.
- Waste in transportation pipeline-related pits.
- Caustic or acid cleaners.
- Boiler cleaning wastes.
- Boiler refractory bricks.
- Incinerator ash.
- Laboratory wastes.
- Sanitary wastes.

These lists are not all-inclusive, and determinations still must be made as to the exempt or nonexempt status of certain other wastes. Table 1-3 lists certain E&P wastes that American Petroleum Institute contends are clearly intended to be exempt under EPA's regulatory determination but were not listed. Operators and contractors should seriously consider testing nonexempt wastes whenever there is reason to believe such wastes may exhibit any of the hazardous waste characteristics.

TABLE 1-3

Additional Exempt Exploration and Production Wastes

- Excess cement slurries and cement cuttings.
- Sulfur contaminated soil or sulfur waste from sulfur recovery units.
- Gas plant sweetening unit catalyst.
- Produced water contaminated soil.
- Wastes from the reclamation of tank bottoms and emulsions when generated at a production location.
- Production facility sweetening and dehydration wastes.
- Pigging wastes from producer operated gathering lines.
- Production line hydrotest/preserving fluids utilizing produced water.
- Iron sulfide.

SOURCE: AMERICAN PETROLEUM INSTITUTE.

Having a complete understanding of RCRA and its application to the drilling and production industry is imperative. Not understanding it can be costly. If a contractor or operator mixes waste listed under Subtitle C with waste from Subtitle D, regardless of the amount of Subtitle C waste, the entire mixture will be considered hazardous waste and must be handled under Subtitle C regulations. For example, dumping pesticides in a reserve pit would render the entire volume of drilling muds, cuttings, rigwash, excess cement, and benign completion fluids reclassified as hazardous waste, despite their otherwise harmless nature.

Attempts to violate Subtitle C regulations can result in harsh penalties by the EPA. Any person or company found violating Subtitle C regulations can be liable for civil penalties of as much as $25,000/day of violation. Furthermore, RCRA also can impose criminal liability of as much as $50,000 and 2 years in prison for anyone who knowingly transports hazardous waste without a manifest.

Cercla authorizes the federal government to clean up toxic contaminants at closed or abandoned hazardous waste dumps or sites where the owner or operator is unable or unwilling to perform such a cleanup when there is a release or threatened release of hazardous or toxic substances. Cercla allows the government to recover costs of cleanups by bringing suit against any parties that may have contributed to the hazardous or toxic material dump.

Under Cercla, the government attempts to locate all potentially responsible parties that may have contributed to waste at the Superfund site. Superfund money is utilized when the emergency actions are found to be necessary or when corrective actions have failed to be implemented. If the government can identify only one party that contributed to the Superfund site, that party alone may be responsible for the total cost to clean up the site. If the government fails to locate any more contributors of waste to the site, it will be the identified party's responsibility to locate and identify other contributors to share in the cost of the cleanup. This is why it is extremely important that research of the site where the waste is to be sent is conducted prior to sending waste there. The contractor or operator should investigate and ensure that the site is approved by EPA and that there are enough funds to properly close the site in the event it is mandated by the government if the site can no longer sustain operations for financial reasons.

SARA is designed to help inform communities of hazardous chemicals produced or released at facilities. The program was developed to ensure proper response in the event of accidental releases of hazardous or toxic chemicals in order to avoid a potential Bhopal-type disaster.

SARA is made up of four parts: emergency planning, emergency release notification, the community right to know and hazardous chemical reporting requirements, and toxic chemical release reporting and emission inventory. The reporting requirements are the same as those under Cercla.

The Clean Water Act of 1977 and the Water Quality Act of 1987 address federal water pollution control and requires the National Pollutant Discharge Elimination System (Npdes) discharge permits. Basically these acts control the discharge of effluent from their point of origin in federal waters. The Npdes outlines specific discharge limits based on most recent technology, water quality standards, compliance schedules for the installation of new pollution control equipment, monitoring requirements and reporting requirements.

Various sections of the these acts prohibit the discharge of oil or hazardous substances in harmful quantities into federal waters, provide for the cleanup of the discharge, and require that companies prepare spill prevention, control and countermeasure (SPCC) plans. The Npdes also requires that proper reporting of discharges be made by the contractors or operators. On most oil and gas operations, a stormwater runoff permit is required. This governs the potential for runoff that may come in contact with machinery or contaminated material onsite that may cause adjacent property contamination. Willful or negligent misconduct of the CWA can result in stiff civil and criminal penalties.

SDWA, which was most recently amended in 1986, is designed to protect all sources of drinking water from contamination. The standards set forth in the act outline the maximum contaminant levels and techniques to safeguard the quality of drinking water supplies.

The SDWA also maintains the underground injection control (UIC) permit program, the sole source aquifer program and the wellhead protection program. The UIC permit program requires that any owner or operator of underground injection wells obtain permits to operate these wells. The permit requires the applicant to prove, through sound and prudent practice and well construction, that underground injection will not contaminate drinking water sources.

The sole source aquifer program states that any aquifer identified as a sole source of drinking water for an area may be classified as a sole source aquifer and that an individual or company involved in the contamination of the aquifer may not receive federal assistance for its cleanup.

The wellhead protection program is designed to allow individual states a means to protect public drinking water from contaminants within their jurisdiction. With this program, the individual states are given the authority to implement programs, including enforcement and penalties, to protect drinking water.

The Clean Air Act (CAA) establishes air quality standards that the individual states must achieve through implementing emissions monitoring programs. The individual states are basically to follow three types of national emissions standards: new source emission standards, and national emission standards for hazardous air pollutants (Neshap), and the prevention of significant deterioration standards (PSD). Under these standards, a contractor or operator is required to offset emissions if a proposed source will have emissions exceeding the new source review levels for nonattainment areas. Although the effects of the CAA do not yet have a direct impact on the drilling side of oil field operations, several changes to the CAA in the near future may require that drilling operations be governed under these regulations.

The National Environmental Policy Act (NEPA) of 1969 requires that all federal agencies use a systematic, interdisciplinary approach for the protection of the human environment. This includes, but is not limited to, use of natural and social sciences in the planning and decision-making processes that may affect the environment. Under NEPA, a detailed Environmental Impact Statement (EIS) is required on any project that may have any adverse impacts that cannot be avoided or mitigated. NEPA is beginning to play a significant role in the drilling of wells in wetland areas, where significant environmental impacts are a concern in such environmentally sensitive areas.

The Endangered Species Act of 1973 establishes a national policy designed to protect and conserve threatened and endangered species and the ecosystems on which they are dependent. Alteration of the natural habitats of endangered species is strictly prohibited under this act. It is the designated agencies' duty to ensure that any proposed action, such as drilling, does not jeopardize the continued existence of a threatened or endangered species and/or results in the adverse modification or destruction of their critical habitat.

2

Preventing Spills

Spill Prevention — Control and Countermeasure Plan

Historical Spills

One way to attempt to comply with environmental rules and regulations governing the oil industry is by instilling philosophies and implementing plans that help minimize the environmental impact of hydrocarbon or other drilling waste spills. Many companies have taken this action and created spill prevention — control and countermeasure plans (SPCC). The SPCC plan basically outlines the steps that the drilling contractor should take to help to avoid spills from occurring and outlines what the contractor should do in the event a spill occurs.

In order to give the reader an understanding of what is contained in an SPCC plan, the following plan is presented in its entirety. If the reader desires more information about developing an SPCC plan or about specifics of a certain SPCC plan, he should contact the governing agency in his area for more information.

Any spills that have occurred prior to the implementation of the SPCC plan have been recorded. The spill record will be kept at the end of this plan. In accordance with EPA regulations, the reports include a written description of the spill, corrective actions taken, and plans for preventing recurrence. Any future reports of spills will also be maintained at the end of the SPCC plan.

Spill Potential

An awareness of possible spill sources is the first step in spill prevention. Spill potential exists in the following situations:

- Tank overflow or leakage.
- Pipeline/flowline rupture or leakage.
- Sump/ring levee/ reserve pit overflow or leakage.
- Loading and transfer of oil/produced water/drilling fluid and cuttings from tank to trucks.
- Storage of empty, used, abandoned and/or unmarked drums.
- Flowing a well unchecked to the atmosphere, especially during drilling and workover operations.

A prediction of the direction of flow from a spill would be shown on attached plats and maps.

Containment

Appropriate containment or diversionary structures are used to prevent discharge materials from reaching waters of the United States as defined by the Code of Federal Regulations (CFR). The following systems may be used for primary containment:

- Ring levees, pit levees, dikes, or retaining walls made of material impervious to oil.
- Culverting, gutters, or other drainage systems.
- Spill diversion or retention ponds.
- Drip/pollution pans.

The following systems may be used for secondary containment:

- Fabric liners.
- Sorbent materials.
- Floating booms.

Tank Truck Loading/Unloading Facilities

During transfer of oil/produced water/drilling fluid and drill cuttings from tank to truck, extreme care and caution must be exercised. Quick coupling connections are used to attach the hose from tank to truck. Sumps are positioned to catch any fluid that may leak during transfer. Outlets and connections are checked for leaks before, during, and after transfer. In addition, all outlets and connections are double valved and plugged when not in use. These procedures assure that no loss or spill results. Sumps must be replaced with above ground containment as soon as possible. It is critical to choose reputable vacuum services and monitor their activities. Hauling trucks should observe speed limits and drive with appropriate caution for existing conditions to avoid any accidents.

Onshore Facility Operations

Drainage

Except when rainwater is being drained, drains from dikes around the tank batteries will be closed and sealed at all times. Prior to draining any rainwater, the fluids inside will be inspected and/or tested to ensure that it complies with applicable water quality standards and will not cause harmful discharge. Any oil or condensate that has accumulated on the rainwater will be removed by vacuum services and returned to a produced water tank or gunbarrel or properly disposed. The drain valve must be opened and resealed, immediately following drainage. All operations are to be conducted under proper supervision and all actions recorded.

Drainage ditches, road ditches, oil traps, sumps and skimmers are all regularly inspected for accumulation of oil. Any accumulations will be properly removed.

Bulk Storage Tanks

Tank material and the construction of tanks used for the storage of oil or condensate must be compatible with the stored fluid and the storage conditions. Each tank battery is to be constructed with a secondary means of containment (dike system) sufficient to contain the volume of the largest tank and any stormwater that may be present. The dikes are made of nonpermeable material. All dikes are equipped with a stormwater drain valve and drained as described in the preceding section.

Buried and partially buried metallic storage tanks represent a potential for undetected spills and therefore should be removed as soon as possible. New buried tanks will not be installed unless absolutely necessary. If new buried tanks are installed, they must meet EPA requirements.

Above ground tanks, tank supports, and foundations will be periodically inspected for mechanical integrity and deterioration for leaks that might cause a spill or accumulation of oil inside diked areas.

The capacity of most tanks is to be monitored through the use of electronic leveling sensing devices. The person in charge of each facility will determine when tanks are near capacity and orders the trucking company to pick up a load of oil. The sensing devices are calibrated and maintained on a regular basis.

Oil and condensate tanks should be equipped with overflow equalization lines to reduce overflow potential. As a tank nears capacity, these overflow lines allow fluid to be diverted to an adjacent tank. In many cases, the overflow lines are piped to emergency tanks. Adequate vacuum protection to prevent tank collapse during a run must be available.

Facility Transfer Operations

Above ground valves and pipelines are to be examined periodically on a regular basis for general conditions of flange joints, valve glands and bodies, bleed valves, etc. Saltwater disposal facilities must be inspected regularly to detect possible system upsets that could cause an oil discharge.

Pipelines are to be protected against corrosion by coating and wrapping, painting, or in some cases by cathodic protection systems. In areas where produced fluids are highly corrosive, chemical corrosion inhibitors must be injected into the flowline system. Each facility must have a continuing flowline maintenance program to prevent spills. Lines are to be replaced or repaired as required. In addition, pipeline and flowline supports must be designed to minimize abrasion and allow for expansion and/or contraction of the pipe.

Drilling and Workover Operations

Mobile drilling and workover equipment must be located in a manner to prevent a spill from reaching waters of the United States as defined by the CFR. Fabric, ring levees, pits, diversion structures, booms, sorbent pads, and tanks are to be used when necessary to control and contain spills of fuel, crude oil and/or drilling fluids.

In addition, other pollution prevention measures, design and operation as outlined in the regulations are to be used to minimize potential spills. After drilling or workover operations are completed, containment devices must be emptied and removed. The fluid from containment devices is disposed of in compliance with local, state, and federal regulations.

Blowout prevention equipment (BOPE) capable of controlling the maximum expected wellhead pressure is to be installed and tested before drilling below any casing string and as required during workover operations. BOPE installations should conform to local, state, and federal regulations. Casing must be designed, run, and cemented in accordance with local, state, and federal regulations and good engineering practices.

PREVENTIVE MAINTENANCE

It should be a fundamental part of each company's policy to make all reasonable efforts to prevent spills. Field operations must be shut down if a natural catastrophe such as a hurricane or tornado is imminent.

A preventive maintenance schedule should be set up for each facility. During preventive maintenance inspections of separation facilities, flowlines, secondary containment and monitoring equipment are required to ensure that spills do not occur. Machinery such as compressors, separators, pumps, and line heaters are to be examined to ensure proper and efficient operation. Any worn or broken parts must be replaced at this time. Preventive maintenance also entails replacing segments of pipelines or flowlines that are old, are in poor condition, or have a history of leakage. A record of all maintenance procedures is crucial.

Inspections

Periodic inspection of tank batteries and other facilities is necessary to protect against spills. During an inspection the following items will be verified:

- Valves are in good working condition and function correctly.
- Tanks show no sign of corrosion or leaks.
- Dikes are maintained at specific dimensions for containment and are free of cracks, debris and weed growth.
- Diked areas are empty of oil, water, and oil residue.
- Sumps show no visible signs of leaks, cracks or overflow.
- A written record of such inspections is kept to log the condition of each facility and any potential for mechanical failure.
- The current SPCC plan should be accurate and true. If not, an SPCC plan revision form will be completed by the person in charge and sent to the appropriate regulatory compliance and/or environmental department.

Training

All personnel are required to attend a training course covering SPCC actions, requirements and goals. The following is an outline of a potential SPCC training program:

- Harmful effects of spilled oil, hazardous substances or produced water.
    - On various land and ecological areas.
    - On wildlife and humans.
    - On water sources and aquatic life.
    - Personnel protective equipment requirements.
- Regulations.
    - Federal.
    - State.
    - Local.

- The SPCC plan.
    - Prevention.
    - Reporting Procedures.
    - Spill Control Techniques.
    - Spill Cleanup Procedures.

        Onshore — confined to land.

        Onshore — reaches water source.

        Inland waters.

        Bay waters.

        Special considerations relating to hazardous substances spills.

In addition to a general training program, field personnel must be aware of all site specific SPCC information pertaining to the area in which they are operating, including:

- Environmentally sensitive areas.
- Inventory, location and operation of onsite equipment.
- Facility layout and operation.

Each field location should stage spill drills annually to provide spill scenarios to test emergency response time, procedures, and effectiveness.

3

SITE ASSESSMENTS

It is inevitable that some functions in drilling operations will directly or indirectly cause some measurable environmental effects. During the drilling of even the most environmentally conscientious well, some degree of environmental dislocation will occur.

For instance, obtaining access to a location often requires that a road be built. This means that after mitigating the possible crop or tree damage with the land owner(s), tens to thousands of trees or shrubs will need to be cut in order to place a board, limestone, or other natural material road. Depending on the remoteness of the well, this could be just a few feet or tens of miles of road.

Before any remediation project is to be undertaken however, it is important to understand what level of remediation a particular site must undergo. This is why it is crucial that proper investigation of the condition of the property should first be conducted through a thorough Phase I environmental site assessment (ESA) and pre-site background sampling. It is worth mentioning the importance of obtaining pre-site or background samples of the area to determine what condition the land was in prior to commencing operations. By not knowing pre-site or background conditions, hundreds of thousands of dollars could be spent on trying to attain a level of remediation that perhaps is nonattainable in that particular area.

One way that a drilling contractor or operating company can minimize environmental liabilities is by conducting a complete Phase I ESA on the property prior to breaking ground for the operation. If conducted properly, the Phase I ESA will identify any areas of concern or of environmental impairment on the property and can supply the drilling contractor or operating company with information about existing conditions in the event of a dispute between the land owner or regulatory agency as to the environmental condition of the land once drilling or production on the property is completed. This becomes an extremely sensitive issue when the land owner does not own the mineral rights for the acreage being utilized for the drilling operations or if the well is found not to be commercially productive.

Phase I Environmental Site Assessment

To ensure that a particular site has been thoroughly environmentally investigated, the Phase I ESA should contain at least the following criteria, which are outlined in more detail in the American Society for Testing and Materials (ASTM) publication E1527, "Practices for Environmental Site Assessments: Phase I Environmental Site Assessment Process":

Site Location and Description

- Legal description including section, township and range; lot, block, and subdivision; and municipality, county, state, and zip code.
- Property type (i. e. , commercial, residential, farm land, undeveloped land, etc.), size of property, and any structures.
- Historical use of the property for at least 40 years back in 5 year intervals.

Surrounding area review

- Review surrounding properties for past and current land usage that may adversely affect the subject site.

Site History

- Conduct a title search on the property for at least 50 years and obtain a title abstract. Report any conditions that may contribute to the current state of the property.
- Obtain aerial photographs of the property for at least the last 40 years in 15 year intervals. The interval should be determined by the change in site activity or site development.
- If structures are present at the site, obtain blueprints or construction documents of these structures.
- Obtain maps or other data regarding geologic, hydrogeologic, and coastal zone conditions at the site and at the property surrounding the site.
- Review and interpret special resource data provided by governmental agencies such as:

  National Park Service.

  National Wilderness Preservation System.

  Environmental Protection Agency.

  U.S. Fish and Wildlife Service.
- Review and interpret federal, state, and local public records to assess the site as well as areas surround the site including:

  City engineering plans.

  Property tax records and appraisal records.

  Zoning/land use records.

  Local and State trial court records.

  Historical/archaeological records.
- Review Compliance Records including:

  National Priority List.

  Comprehensive Environmental Response, Compensation and Liability Information System List.

  Resource Conservation and Recovery Act.

  TSD Facilities List.

  Emergency Response Notification System.

  State lists including but not limited to:

Underground Storage Tanks.

Leaking Underground Storage Tanks.

Solid Waste Facilities.

- Interviews.

  Interviews of any available individual familiar with the site or adjacent to the site to determine the historical land use activities at the site should be conducted. These should include past owners, surrounding land owners, governmental agencies, real estate sales persons, appraisers, etc.

Site Investigation

- Visual Inspections.

  A visual inspection, including obtaining a photographic log of the site, should be conducted. Any special resources existing on the site should be noted at this time (crops, water, etc.).

  A visual inspection of the surrounding area should also be conducted.

- Investigate for environmental hazards by investigating:

  The topography of the site and surrounding sites.

  The geology of the site and surrounding sites.

  The hydrology of the site and surrounding sites

  The use of chemicals and/or raw materials at the site or surrounding sites.

  The presence of polychlorinated biphenyls (PCBs) or equipment potentially containing PCBs located on or adjacent to the site.

  The potential for asbestos-containing materials on the site or surrounding sites.

  The presence or potential of radon on the site or surrounding sites.

  The presence of any other hazardous substances on the site or surrounding sites.

  The condition of the groundwater and/or drinking water at the site or surrounding sites.

The current or past presence of any landfills, pits, sumps, or dry wells on the site or surrounding sites that may potentially contribute to contaminating groundwater or soil.

The presence of any storage tanks that may have had a discharge and potentially caused groundwater or soil contamination at the site or surrounding sites.

The stormwater drainage pattern of the property and surrounding properties.

The presence of wetlands to determine if the site or surrounding sites are in a wetlands area.

If in the Phase I ESA it has been determined that there is the likelihood of environmental impairment on the property, then a detailed sampling and analysis plan should be drafted for the area. The sampling and analysis plan should detail the review of the suspected impairment, outline a plan to comprehensively sample the area(s) of suspected contamination to delineate the extent of contamination, and detail the type of laboratory analysis that should be run to confirm the type of contamination.

Pre-Site Sampling

Once a comprehensive sampling and analysis plan has been drafted and reviewed, the pre-site sampling should be conducted.

There are several methods of determining the pre-site conditions. The most popular and probably the most economical is that of taking soil samples in and around the area designated for drilling operations.

Knowing the chemicals that will be used during drilling operations is extremely important in determining the type of pre and post-sampling that should be conducted at the location. In other words, if you are going to use a barite drilling fluid system, the contractor should test the site for barium prior to the operation and after the operation is complete.

It is equally important that the drilling contractor know the background conditions of the site prior to operation commencement. If an area is naturally high in cadmium, then when the post-site sampling and testing indicates that the site is high in cadmium, the contractor will not spend potentially tens of thousands of dollars in trying to bring down these higher values of cadmium to levels that just cannot be attained.

The EPA has given individual states jurisdiction over the oil and gas industry to determine the constituents and their acceptable levels that drilling and production locations must adhere to.

What this basically means is that there is no set standard that the industry can utilize to determine what should be tested for in a pre or post-site assessment of the property.

In reviewing the guidelines from every state that has drilling operations, it was determined that the Louisiana Department of Natural Resources Office of Conservation maintains the most stringent parameters that drilling sites must meet after drilling operations are concluded. These parameters are contained in Louisiana Statewide Order 29-B. This statute has become something of an international standard for environ-

TABLE 3-1

LOUISIANA ORDER 29-BTEST PARAMETERS

TEST PARAMETER	UPLAND CRITERIA	MARSH CRITERIA	SOLIDIFICATION
pH	6-9	6-9	6-12
Arsenic	10 mg/l	10 ppm	<0. 5 mg/l
True Total Barium	40,000 mg/l	20,000 ppm	<10. 0 mg/l
Cadmium	10 mg/l	10 ppm	0. 1 mg/l
Chromium	500 mg/l	500 ppm	<0. 5 mg/l
Lead	500 mg/l	500 ppm	<0. 5 mg/l
Mercury	10 mg/l	10 ppm	<0. 02 mg/l
Selenium	10 mg/l	10 ppm	<0. 1 mg/l
Silver	200 mg/l	200 mg/l	<0. 5 mg/l
Zinc	500 mg/l	500 mg/l	<5. 0 mg/l
Oil & Grease	<1% dry wt.	<1% dry wt.	<10. 0 mg/l
Electrical Conductivity	<4 mmhos/cm	<8 mmhos/cm	NA
Sodium Absorption Ratio	<12	<14	NA
Exchangeable Sodium Percent	<15%	<25%	NA
Chlorides	NA	NA	<500. 00 mg/l

mental standards for drillsites. Therefore it is recommended that prior to any drilling operation is commenced, the soil is tested for the parameters contained in Table 3-1.

As is shown in this chart, the criteria differ for uplands, wetlands and solidification operations. Being that wetlands potentially contain more sensitive plant and animal life than uplands, the criteria are more stringent than for uplands. Being that solidification (i.e., a trench is made and the material buried) is only allowed in upland areas, the criteria are even more lenient than for wetlands areas.

Two constituents that should definitely be tested for in a pre-site assessment of a drilling operation are barium and oil and grease. These are extremely important due to the fact that barite is used on almost every drilling well as a weighting agent, and with all of the moving parts on a rig that require lubrication, the chance of oil and grease spillage will be greater.

If the site is located in an area that will eventually be farmed, special attention should be paid to the electrical conductivity of the soil. Although some crops can tolerate elevated levels of electrical conductivity, other crops, such as sugar cane, cannot. Electrical conductivity is a function of the salt content in the soil.

On drilling sites where lead-based pipe compound is used, it is important that the area around the well cellar be tested for concentrations of lead. The lead is easily transported from the rig floor to the ground by washing down the rig floor for housekeeping and safety purposes.

Once the drilling operation is complete and the turnaround removed, the drilling contractor and/or operating company should always conduct post-site sampling of the site for at least the same constituents that were conducted in the pre-site assessment of the property. If it is determined that other potential constituents may have been introduced during the drilling operation, the company should test for these as well. If it is found that elevated levels of any constituent is found in the post-assessment sampling that exceed the pre-assessment sampling, preparations should be made to have the area remediated.

SAMPLING

SAMPLING TECHNIQUES

When conducted properly, sampling can provide a good representation of the composition of the soil or groundwater at a particular site. If conducted improperly, however, a contractor could spend countless hours and tens of thousands of dollars attempting to remediate a site that has no contamination.

Because sampling is such an important part of a pre-site and post-site assessment, it is important to follow proper protocol to ensure that representative samples are being obtained. Furthermore, if sampling is conducted improperly, cross contamination from a contaminated zone to an uncontaminated zone could occur.

The single most important factor in conducting any sampling event is that all of the sampling equipment be clean and free from contamination. Obtaining samples with tools tainted with contamination from an outside source could cause a remediation to be recommended when there is no contamination or for an improper remediation technique to be applied at a site because the wrong contaminant has been identified as the culprit.

There are three basic rules that should be followed when sampling a location. First, always decontaminate the sampling equipment prior to commencing any sampling event. Second, always decontaminate the sampling equipment between boreholes to prevent hole-to-hole contamination. Third, never place a clean piece of sampling equipment on the ground so that surface soil will contaminate the equipment. Although this may seem simple in concept, when the temperature is 110° F and the sampler has been on the job for 10 hr, sometimes shortcuts are taken to expedite the process.

There are several methods utilized today to obtain samples. Soil samples on shallow, non-groundwater penetrating sampling events can be taken with a spoon, with ordinary hand soil augers and with soil probes. These types of samplers are utilized mainly to obtain shallow samples on pit closures or site closures. On these types of situations, if the equipment is decontaminated properly, there is little to no chance any cross contamination will occur. It is usually not necessary to have a drilling rig to obtain these samples. These can usually be taken by hand.

When groundwater penetration is anticipated, however, sampling and the techniques for sampling become more critical. It is not appropriate to use standard sampling equipment when penetrating groundwater. Since the standard sampling equipment is not designed to transport cuttings out of the sample hole, the potential for dragging contamination down from soil to water, soil to soil, or water to soil increases substantially. That calls for using continuous flighted augers at the very least.

Unlike the basic auger sampler, the continuous flighted auger will carry cuttings from the hole to the surface, thus minimizing the potential for contamination being dragged downhole. Because these types of samples are usually deeper, a drilling rig is usually utilized to take these samples.

When it is critical to obtain undisturbed soil samples, hollow flighted augers should be used. Like the continuous auger system, the hollow auger carries the cuttings from the hole to the surface as they are being cut. The exception is that undisturbed soil and water samples can be taken during the drilling of the boreholes without having to remove the drill stem from the well. This again minimizes the chance for contamination to be dragged downhole and expedites the sampling event. This technique requires a drilling rig to be used to take the samples.

Sample Analysis

After conducting the initial site assessment of the prospective drilling site and taking samples, it is important to understand the types of laboratory analyses that are required to properly assess the samples for the most likely contaminants. In a drilling operation, definition of the chemicals to be used is fairly straightforward. For instance, on any given well, drilling fluid chemicals such as barite, bentonite, caustic soda, and lignosulfonate are likely to be used. Therefore, it is necessary to conduct laboratory analyses to check for the presence of these chemicals in the soil or groundwater.

Other functions during drilling operations also can contribute to other forms of contamination on a site. For instance, hundreds of pounds of thread compound are utilized during drilling operations. In order for thread compound to be effective, zinc and other metals are part of the makeup of the thread compound. Because hundreds of connections are made on each well, and it is important to keep the rig floor clean for safety reasons, high concentrations of zinc tend to accumulate around the wellbay area. So it becomes necessary to test the soil or water for zinc.

Most states have adopted guidelines as to the type of analyses that are required at a drill site. For a typical drilling operation, the required analyses are shown in Table 4-1.

TABLE 4-1

DRILLING SITE SAMPLE ANALYSIS PARAMETERS

- pH value
- Arsenic
- True total barium
- Cadmium
- Chromium
- Lead
- Mercury
- Selenium
- Silver
- Zinc
- Oil and grease
- Electrical conductivity
- Sodium absorption ratio
- Exchangeable sodium percentage

The pH (alkalinity/acidity) of any medium can be affected greatly by just about any chemical. The pH value should range between a 6 and a 9 to be considered neutral. Caustic soda beads are usually added to the drilling fluid to enhance the reaction of other chemicals used. The addition of caustic usually will increase the pH value.

The metals in the soil or water are usually affected by the various chemicals utilized in the drilling fluid or by other chemicals utilized on the rig. The allowable limits are usually set by EPA.

Oil and grease levels can vary widely on any given rig. On purely mechanical drilling rigs, massive amounts of oil and grease are utilized to run and lubricate each moving part. Unfortunately, as with most mechanical moving parts, leaks occur. The limits in most states is 1% for oil and grease.

The electrical conductivity (EC), sodium absorption ratio (SAR) and the exchangeable sodium percentage (ESP) are all functions of salinity. This is usually affected by the drilling fluid make-up water which is either trucked, taken from the nearest body of water, and/or is supplied by a water well on location.

For many environmental laboratories, the running of samples for these analyses is standardized. So the presentation of the findings from the analyses should be in a form that is easily understood. But as with any service, it is essential to always make sure that the analyses are accurate. This is done through a proper quality assurance/quality control (QA/QC) program. Although the laboratory may have its own QA/QC program, it is important that the contractor or oil company insure against fraud or misinterpretation. Misrepresentation of data could cost the operator or contractor thousands of dollars if left uncorrected.

There are several steps in initiating and implementing a QA/QC program. At a minimum, the following guidelines should be followed in any QA/QC plan. First, focus on identifying the end use of the data and on determining the degree of precision, representation of accuracy, completeness, and comparability necessary to satisfy its intended use. Next, field-generated quality control samples should be collected during the soil sampling program. Field-generated QA/QC samples should include field duplicates, trip blanks, and rinsate blanks.

Prior to spending any money on having samples laboratory tested, it is important to identify what the analyses are intended for. By not having an understanding of the intended use of the data, many unnecessary and costly tests often are run. By outlining the objective of the analyses, a laboratory analysis program can be tailored to the needs of the project. The operator can save thousands of dollars by running tests that are needed rather than those that are convenient.

The field duplicates should be collected at a frequency of 10-15%. This will allow you to check the repeatability of the laboratory equipment. Trip blanks should be collected for each batch of samples sent in to the laboratory. Trip blanks should never enter the sampling area during sample collection but should be included inside the coolers prior to packaging for delivery to the laboratory. In addition, the trip blank is not to be opened prior to reaching the laboratory. Trip blanks will assist in determining if "boilerhousing" by the laboratory is going on.

Rinsate blanks should be taken at a 10% frequency per type of sample collection equipment. A rinsate blank should be collected after standard decontamination of equipment by fully rinsing with deionized water all field equipment that normally would contact the sample. The rinsate water also should rinse any equipment used in the homogenization of field duplicates. The rinsate blank is intended to ensure that proper decontamination procedures are being implemented in the field and cross contamination is being kept to a minimum.

Data Interpretation

Once the samples have been collected and analyzed by the laboratory, it is important to properly interpret the results. Even the best and most expensive test results are only as good as their interpretation. Understand the detection limits of the instruments analyzing the sample. If a sample indicates "non-detect" for a certain analyte and the detection limit of the equipment is 10 ppm, but it is critical to know if any of the analyte is present, then either the detection limit should be lowered or a different test conducted to determine the analyte concentration. Understanding the limits of the laboratory analyses also is extremely important if plans call for mixing and diluting soil in the field as a means of remediation. If the concentrations and quantity of the contaminants are too high, then fresh, non-contaminated soil may need to be brought in to assist in the dilution process or in the worst case a portion or all of the contaminated soil may have to be hauled off to remediate the site.

5

WELLSITE DETERMINATIONS

The determination of whether or not a proposed surface drilling location is in an environmentally sensitive location has been a growing concern to operators and drilling contractors since the late 1980s. The determination that a site is in a sensitive wetlands area or in an endangered species area can mean at the least several months of delay in drilling because of the amount of permitting that first must be conducted. When determining the time value of money, this can be an extremely costly wait. Further, locating a drilling site in an environmentally sensitive area can invite objections from environmentalist groups or the public at large. That, in turn, can lead to costly litigation or permit denials that also must figure into project economics.

Determining if an area that is to be the site of drilling operations is in an environmentally sensitive area has become a science. Deciding where the site is located and what type of vegetation and soil type will be affected contributes to the determination of whether a site is considered a wetlands or an uplands site. Depending upon which state the well is located in, a state or federal government agency will have the final decision as to whether the site is situated in an environmentally sensitive area.

WETLANDS

One area of environmental concern that has gained considerable notoriety in recent years is that of wetlands. Although once thought of as vast wasteland with little intrinsic value or potential to add value, recent ecological studies have indicated that wetlands serve an extremely important role in the preservation of many species of plant and animal life and perform a vital function in preserving the natural balance of larger ecosystems.

Wetlands are fragile ecosystems that perform invaluable functions such as:

- Water quality enhancement. Wetlands remove pollution from waters that flow through them — in effect, functioning as biological sewage treatment plants.
- Physical protection. Wetlands protect shorelines from wave or storm erosion and protect downstream areas from damaging effects of floods by slowing and temporarily storing floodwaters.
- Climatic influences. Wetlands participate in the global cycles of nitrogen, sulfur, methane, and carbon dioxide and may help control atmospheric pollution by removing excess nitrogen and CO2 produced by human activity.
- Water supply. Some wetlands act as groundwater recharge zones for area aquifers. Many store water during wet parts of the year and release it at relatively constant rates, helping to maintain regular stream flows.
- Wildlife habitat. Many species of fish and wildlife, including a large proportion of federally listed threatened or endangered animals (45%) and plants (26%), depend directly or indirectly on wetlands to provide critical breeding, nesting, rearing and wintering habitat.
- Food chain support. Coastal and riverine wetlands produce large quantities of food materials that are exported to estuaries and other coastal areas where they support marine food webs, many of which are critical to commercial fisheries.
- Commercial products. Wetlands serve as important sources of fish and shellfish, timber, forage, wild rice, cranberries, blueberries, animal furs and other useful materials.
- Recreation and aesthetics. Wetlands provide places for hunting, fishing, nature study, boating, and outdoor education and are valuable simply for their natural beauty.

Whether the agency making the determination is a state or federal agency, it must follow EPA guidelines governing wetlands areas. These regulations can be found in the Clean Water Act under Section 404. Under Section 404, a permit is required from the U.S. Army Corps of Engineers before any dredge or fill materials can be discharged into federal waters. If a proposed site does not meet the specifications for a wetlands area, the site will be considered a non-wetlands site and thus no special permitting will be required to build the drilling location and drill the well.

When most people think of wetlands, they think of Louisiana swampland. In reality, wetlands can exist in all areas of the country. According to EPA, wetlands have very little to do with elevation and can be found in practically every county or parish in the nation. The determination of any area as a wetlands has become a complicated and often controversial process. There are basically ten types of wetlands: inland freshwater marshes, inland saline marshes, bogs, tundra, shrub swamps, wooded swamps, bottom lands, coastal salt marshes, mangrove swamps and tidal freshwater wetlands (Table 5-1). The primary factors distinguishing wetlands are: location (whether the land is coastal or inland), salinity (freshwater or saltwater), and vegetation (swamp, marsh, or bog) (Figure 5-1).

Inland freshwater marshes may occur at any latitude but are not common at very high altitudes. Their water depths generally range from

FIGURE 5-1

WETLANDS COMPARISONS

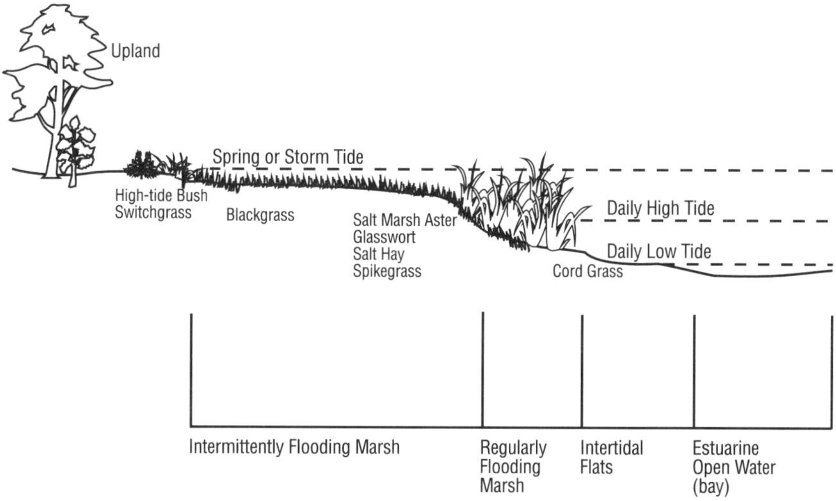

TABLE 5-1

WETLAND TYPES

Inland freshwater marsh	Dakota-Minnesota drift and lake bed; upper Midwest; and Gulf coastal flats	North Dakota, South Dakota, Nebraska, Minnesota, Florida
Inland saline marshes	Intermontane; Pacific mountains	Oregon, Nevada, Utah, California
Bogs	Upper Midwest; Gulf-Atlantic rolling plain; Gulf coastal flat; and Atlantic coastal flats	Wisconsin, Minnesota, Michigan, Maine, Florida, North Carolina
Tundra	Central highland and basin; arctic lowland; and Pacific mountains	Alaska
Shrub swamps	Upper Midwest; Gulf coastal flats	Minnesota, Wisconsin, Michigan, Florida, Georgia, South Carolina, North Carolina, Louisiana
Wooded Swamps	Upper Midwest; Gulf coastal flats; Atlantic coastal flats; and lower Mississippi alluvial plain	Minnesota, Wisconsin, Michigan, Florida, Georgia, South Carolina, North Carolina, Louisiana
Bottom land hardwood	Lower Mississippi alluvial plain; Atlantic coastal flats; Gulf-Atlantic rolling plain and Gulf coastal flats	Louisiana, Mississippi, Arkansas, Missouri, Tennessee, Alabama, Florida, Georgia, South Carolina, North Carolina, Texas
Coastal salt marshes	Atlantic coastal zone; Gulf coastal zone; Eastern highlands; Pacific mountains	All Coastal States, but particularly the Mid-and South Atlantic and Gulf Coast States
Mangrove swamps	Gulf coastal zone	Florida and Louisiana
Tidal freshwater wetlands	Atlantic coastal zone and flats; Gulf coastal zone and flats	Louisiana, Texas, North Carolina, Virginia, Maryland, Delaware, New Jersey, Georgia, South Carolina

6 in. to 3 ft. Marsh vegetation is characterized by soft-stemmed plants, grasses, sedges, and rushes that emerge above the surface of the marsh. They include such common plants as water lilies, cattails, reeds, arrowheads, pickerel weed, smartweed, and wild rice.

Inland saline wetlands occur primarily in shallow lake basins in the western United States. They usually are saturated during the growing season and often covered with as much as 2-3 ft of water. Vegetation is mainly alkali or hard-stemmed bulrushes, often with wideon grass or sago pondweed in more open areas.

Bogs occur mostly in shallow lake basins, on flat uplands, and along sluggish streams. The soil, which often consists of thick peat deposits, usually is saturated and supports a spongy covering of mosses. Woody and/or herbaceous vegetation may also grow in bogs. In the northern U.S., leather leaf, Labrador tea, cranberries, and cotton grass often are present. Cyrilla, persea, gordonia, sweetbay, pond pine, Virginia chain fern, and pitcher plants grow in southern bogs, which are found on the southeastern coastal plains. These bogs are more commonly known as "pocosins."

A tundra is essentially a wet arctic grassland dominated by lichens (reindeer moss), sphagnum mosses, grasses, sedges, and dwarf woody plants. It is characterized by a thick spongy mat of living and undecayed dead vegetation that often is saturated with water. Its deeper soil layer or permafrost remains frozen throughout the year. The surface of the tundra is dotted with ponds when not completely frozen. In Alaska, wet tundra occurs at lower elevation, often in conjunction with standing water. Moist tundra occurs on slightly higher ground. An alpine tundra or meadow, similar to the arctic tundra, occurs in high mountains of the temperate zone.

Shrub swamps occur mostly along sluggish streams and occasionally on flood plains. The soil usually is saturated during the growing season and often is covered with as much as 6 in. of water. Vegetation includes alder, willows, button bush, dogwoods, and swamp privet.

Wooded swamps occur mostly along sluggish streams on flood plains, on flat uplands, and in very shallow lake basins. The soil is saturated at least to within a few inches of its surface during the growing season and often is covered with as much as one or two feet of water. In the North, trees included tamarack, white cedar, black spruce, valsam, red maple, and black ash. In the South, water oak, overcup oak, tupelo gum, swamp black gum, and cypress are dominant. In the Northwest, western hemlock, red alder, and willows are common. Northern evergreen swamps usually have a thick ground covering of mosses. Deciduous swamps frequently support beds of duckweeds, smartweeds, and various other herbs.

The areas adjacent to rivers and streams most commonly recognized as bottom land hardwood and flood plain forests in the Eastern and Central United States and as stream bank vegetation in the arid West are called riparian habitats. Riparian ecosystems are unique, owing to their high species diversity, high species densities, and high productivity relative to adjacent areas.

Bottom lands occur throughout the riverine flood plains of the southeastern United States, where more than 100 woody species occur. Bottom lands vary from being permanently saturated or inundated throughout the growing season at the river's edge to being inundated for short periods at a frequency of only 1-10 yr/100 yr at the uplands edge. On the lowest sites that are flooded the longest, most frequently, and to the greatest depths, bald cypress, tupelo gum, button bush, water elm, and swamp privet are most abundant. As elevation increases (and flooding frequency and depth decrease), overcup oak, red maple, water locust, and bitter pecan occur. Nuttal oak, pin oak, sweet gum, and willow oak appear where flooding occurs regularly during the dormant season but where water rarely is present at midsummer. Sites nearest the high-water mark, which are flooded only occasionally, have shagbark hickory, swamp chestnut, oak, and post oak.

Riparian habitats in the arid West are scattered widely along ephemeral, intermittent, and permanent streams that commonly flow through arid or semiarid terrain. Woody vegetation associated with these wetlands includes willows and alders at higher elevations; cottonwoods, willows, and salt cedar at intermediate elevations; salt cedar, mesquite, cottonwoods, and willows at lower elevations.

Salt marshes alternately are inundated and drained by the rise and fall of the tide. Because the plants and animals of the marsh must be able to adjust to the rapid changes in water level, salinity, and temperature caused by tides, only a relatively small number of plants and animals are able to tolerate these conditions. Thus, there is a high degree of similarity in the kinds of species present. Plants of the genus Spartina and the species Juncus and Salicornia are almost universal in their occurrence in the United States salt marshes.

Mangrove is a term denoting any salt-tolerant, intertidal tree species. In the United States, mangroves are limited primarily to Florida coastal areas. Large mangrove swamp forests are found only in south Florida and are especially extensive along the protected southwestern coast. On the northwestern Florida coast, black mangrove occurs mostly as scattered scrublands. On the eastern shore of Florida and along the Louisiana coast, mangroves are found behind barrier islands and on the shores of protected coastlands.

Tidal freshwater marshes occur in virtually every coastal state but are most abundant in the estuaries of the mid-Atlantic coast and along the coasts of Louisiana and Texas. Dominant intertidal plants include a mixture of grasses and broadleaf species, such as arrow arum, spatterdock, pickerel weed, and arrowhead, which form rather complex multilayered plant zones. The upper marsh may have from 20-50 species of grasses, shrubs, ferns and herbaceaous plants. The above described wetland types are distributed unevenly across the United States.

Once it is determined that a site is or contains a wetland area, the regional Corps of Engineers office and appropriate state and local agencies should be contacted for instructions. If at all possible, it is best to forego Clean Water Act jurisdiction by conducting dredge-and-fill activities in upland, nonwetland locations. If activities in wetlands are deemed necessary, it is important that the state water quality or conservation agency is contacted at an early stage. Under Section 401 of the Clean Water Act, state water quality certification is required before a Section 404 permit can be obtained. Any attached water quality conditions are normally made enforceable elements of the Section 404 permits.

There are two types of federal permits for the dredge-and-fill of wetlands: The general permit, which is issued to the public at large on a regional and national basis for standard activities that involve minor work; and the standard permit required for larger projects, which involves public notice and a case-by-case evaluation of proposed activity according to regulatory guidelines.

Under Section 404(f), the following activities are exempt from permitting regulations:

- Normal farming, silviculture and ranching practices (as part of established operations).
- Maintenance, including emergency reconstruction of recently

damaged parts of currently serviceable structures such as dikes, dams, levees, and similar specified structures.

- Construction or maintenance of farm or stock ponds or irrigation ditches or drainage ditch maintenance.
- Construction of temporary sedimentation basins on a construction site that does not include placement of fill material into waters of the United States.
- Construction or maintenance of farm or forest roads or temporary roads for moving farming or mining equipment if best management practices are followed.

These exemptions do not apply if the discharge is part of an activity whose purpose is to convert an area of federal waters into a use to which it was not previously subject.

For activities that require the standard permit, a "pre-application consultation" can give the applicant an idea of the requirements that must be fulfilled in order to construct a drilling site in a wetland location. A standard permit application usually takes 2-6 months to process. Once the application is received, a public notice will be issued, usually within 15 days. Then a 20-30 day comment period follows, after which the application is reviewed by the Army Corps of Engineers. Fees for the application vary from state to state. No fee is charged for a general permit application.

WETLANDS PERMITTING

In reviewing applications (Figure 5-2), the Army Corps of Engineers may rely on the Fish and Wildlife Service for habitat evaluation and other biological issues and EPA for recommendations on water quality and other general environmental issues. But beyond environmental consultation, EPA has an important three-part role in the Section 404 regulatory process: determining Clean Water Act jurisdiction; promulgating and policing implementation of the Section 404(b)(1) guidelines; and enforcing Section 404. In addition, Section 404(c) gives EPA authority to veto permits when development will have an unacceptable adverse effect on municipal water supplies, shellfish beds, and fishery areas (including spawning and breeding areas), wildlife, or recreational areas.

Developed by EPA, the Section 404(b)(1) guidelines (published in 40

CFR Part 230) call for the Corps of Engineers to conduct a "practical alternatives" analysis when determining if a permit will be granted. The analysis hinges on the availability and feasibility of an alternative non-wetland site for the drilling site or project. Alternative siting is vastly preferred to any level of mitigation. There is a presumption that activities that are not water-dependent do not require development in wetlands. The guidelines state that "No discharge of dredged or fill material shall be permitted if there is a practicable alternative to the proposed discharge which would have less adverse impact on the aquatic ecosys-

FIGURE 5-2

CORPS OF ENGINEERS PERMITTING PROCESS

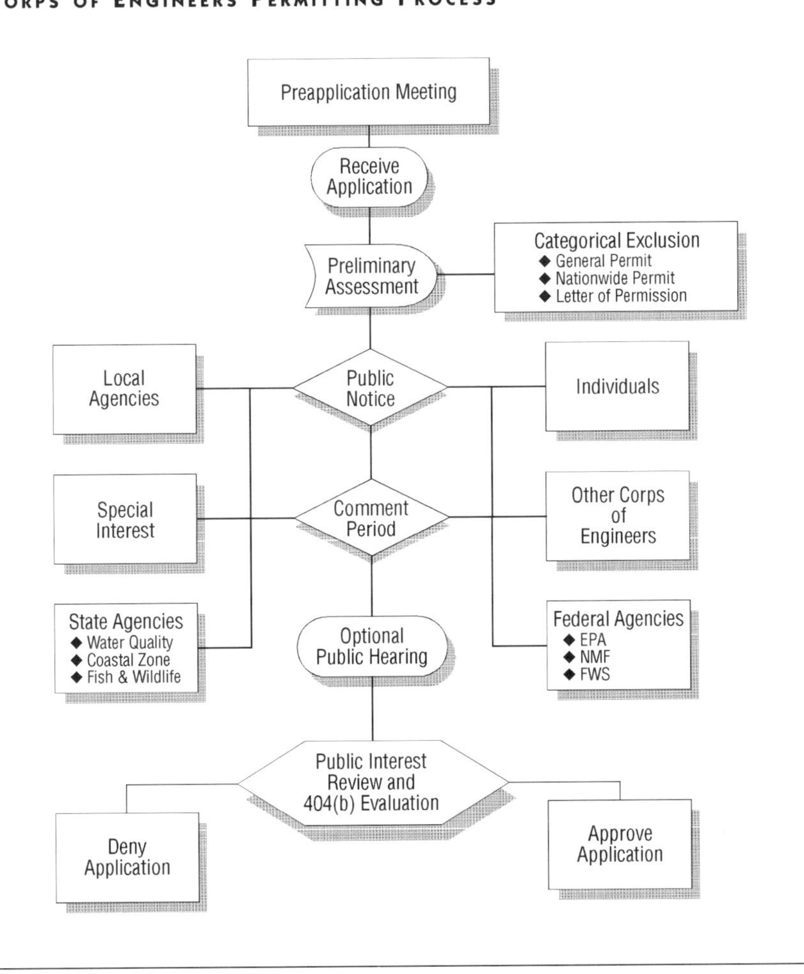

tem, so long as the alternative does not have other significant adverse environmental impacts."

If it is necessary to develop on the wetland site, the major consideration in permit review is whether sufficient mitigation is proposed to offset the adverse environmental effect of the project. There are three ways to mitigate: acquire, enhance and preserve wetlands, restore damaged wetlands, or create new ones. Usually mitigation involves a combination of all three and replacing at least acre for acre and type for type. It is advised to consult with EPA and Fish and Wildlife about acceptable mitigation efforts before submitting a permit application in order to avoid lengthy delays and complications. Much time and expense can be saved by addressing such issues prior to any development commitments. While onsite, in-kind mitigation is normally preferred, off-site, out-of-kind mitigation can at times be quite acceptable and beneficial. A more detailed discussion as to the types of mitigation that are usually required or desired can be found in Chapter 6.

The Corps of Engineers receives as many as 12,000 individual wetland permit applications/year. Recently, only 5% of the applications have been denied. Reportedly this small group consists primarily of applicants who refuse to change the design, timing, or location of proposed activities. Obviously, applicants must be prepared to be flexible and work closely with the Corps of Engineers and EPA to arrive at a development and mitigation plan that meets the needs of all parties.

Wetland regulation occurs under a variety of federal laws in addition to the Clean Water Act: the Coastal Zone Management Act, Estuarine Areas; Water Bank Program for Wetlands Preservation, Rural Environmental Conservation Program, Migratory Bird Conservation Act, Rivers and Harbors Act of 1899, Fish and Wildlife Coordination Act, and Watershed Protection and Flood Prevention Act. Also, there are federal executive orders on wetlands protection and floodplain management. It is of particular note that permits are subject to the provisions of the National Environmental Policy Act (NEPA), which may require environmental assessments and environmental impact studies of the proposed site.

The key to successfully and expeditiously obtaining a permit to drill in a wetlands area is for the drilling contractor or operator to have an understanding of the laws that govern that practice.

Understanding avoids delays and can save hundreds of thousands of dollars in unanticipated or wasted costs. As indicated above, the federal laws governing wetlands usage are numerous. A brief description of each of these laws follows.

The Clean Water Act of 1977

The Clean Water Act bolstered the continuing expansion of the Corps of Engineers' role as wetlands protector. Section 404 of the act prohibits the discharge of dredge or drill material into "navigable waters," defined as waters of the United States, without a permit from the Corps of Engineers. The geographic scope of Section 404 is much broader than Section 10 of the Rivers and Harbors Act of 1899, encompassing more than just traditionally navigable waters.

Section 404 of the Clean Water Act has three main parts:

- Section 404(a) authorizes the Corps of Engineers to issue permits for filling navigable waters, which includes wetlands. The act states that a permit may be issued after notice and opportunity for public hearings for the discharge of dredged or fill material into the navigable waters at specified disposal sites. The act gave the Corps of Engineers authority to issue permits, but no guidelines to evaluate them. So the Corps of Engineers relies heavily on its public interest review to evaluate permits.
- Section 404(b) requires that the Corps of Engineers issues permits in accordance with guidelines developed by EPA. These are referred to as the "b-1 guidelines. " The guidelines state that no discharge of dredged or fill material shall be permitted if there is a practical alternative to the proposed discharge that would have less adverse impact on the aquatic ecosystem. In addition, no discharge of dredged or fill material shall be permitted that will cause or contribute to significant degradation of the waters of the United States.
- Section 404(c) authorizes EPA to veto a decision by the Corps of Engineers to issue a permit to drill in a wetland. It states that it is authorized to deny or restrict the use of any defined area for specification as a disposal site whenever it determines that the discharge will have unacceptable adverse effects on municipal

water supplies, shellfish beds and fishery areas, wildlife, or recreational areas.

- Section 404(e) authorizes the Corps of Engineers to issue general permits on a state, regional, or nationwide basis for certain categories of activities in wetlands that are similar in nature and will cause only minimal adverse effect to the environment.
- Section 404(f) exempts certain activities from the permit requirements such as normal farming silviculture and ranching activities, minor drainage, harvesting for the production of food, fiber and forest products, or upland soil and water conservation practices. Such activities, however, must be part of an established farming, silviculture or ranching operation; otherwise they are not exempt. For example, a farmer many not drain and fill a wetland that has never been farmed without first obtaining a permit from the Corps of Engineers.
- Section 404(g) authorizes states to assume the permit program from the Corps of Engineers provided that the program is approved by EPA.

The Coastal Zone Management Act of 1972

This act provides financial incentives for states to adopt federally approved coastal zone management programs to protect coastal resources, which include beaches, barrier islands, barrier reefs, dunes, and wetlands. Federal actions, such as offshore oil leasing must conform with a federally approved state program. If not, the state may veto the federal action. This is the so-called "consistency requirement," which has been the focus of considerable debate and litigation between the states and the federal government.

Approved state programs must delineate the coastal zone boundary, indicate which activities are permissible within the defined coastal zone, inventory special resource areas requiring protection, establish a policy framework to guide decisions about appropriate resource use and protection, and include sufficient legal authority to implement the program.

Twenty four of the 30 coastal states, including the Great Lakes states, have federally approved coastal zone management programs.

The Coastal Barrier Resources Act of 1982

This act restricts, and in some cases eliminates, federal subsidies for building on undeveloped coastal barriers. The act does not prohibit development on coastal barriers, but it does prohibit federal expenditures and financial assistance, such as federal flood insurance, for such development.

The National Flood Insurance Act of 1968

This act provides financial incentives for communities to adopt federally approved floodplain management programs. Administered by the Federal Emergency Management Agency (FEMA), the program utilizes a financial carrot and stick approach to coax communities into adopting programs that will ultimately reduce the loss of lives and property from floods. For communities with approved programs, the federal government provides subsidized flood insurance to those who own property in the floodplain. Communities that do not participate in a program to regulate future floodplain uses are ineligible for federal disaster assistance. In general, the programs apply to new and rebuilt construction in floodplains and usually include restrictions on the type and location of development. Although not its primary focus, the program covers development in wetlands, since nearly all coastal and most inland wetlands occur in floodplains.

The Endangered Species Act of 1973

This act was enacted to protect rare plants and animals, such as the California condor, that are in danger of becoming extinct. The act requires federal agencies, in consultation with the U.S. Fish and Wildlife Service and the National Marine Fisheries Service, to ensure that any action authorized will not jeopardize endangered or threatened species directly, nor hurt or destroy their habitat, including wetlands. It also prohibits any person from taking an endangered species. Taking an endangered species includes hunting, trapping, harming, or harassing such species.

Fish and Wildlife Coordination Act of 1934

This act requires the U.S. Army Corps of Engineers to consider the comments of federal and state fish and wildlife agencies, such as the United States Fish and Wildlife Service or the National Marine Fisheries Service, before issuing a Section 404 permit.

The National Environmental Policy Act of 1969

In December 1969, Congress passed the National Environmental Policy Act (NEPA) and ushered in what is often called "the environmental decade." The act reflected growing concern throughout the United States that unfettered economic growth was spoiling the water, air and land on which all life depends. NEPA was enacted to reconcile conflicts between economic growth and environmental protection. It directs all federal agencies to consider the impacts of major federal actions on the environment. NEPA does not prohibit development in environmentally sensitive areas but requires all federal agencies, in making decisions about federal or federally permitted projects — including private projects requiring federal permits — to consider environmental impacts of a proposed federal action. Section 102(2) of NEPA states that all agencies of the federal government shall insure that presently unquantified environmental amenities and values be given appropriate consideration in decision-making along with economic and technical considerations.

NEPA created the Council on Environmental Quality as an agency in the Executive Office of the President. The Council on Environmental Quality was originally created to coordinate federal compliance with NEPA, but since Congress did not grant it authority to adopt regulations, the Council on Environmental Quality at first exercised only an advisory role. In 1978, however, it was granted authority to issue regulations that provide an interpretation of NEPA and that establish uniform procedures for preparing an environmental impact statement (EIS) or an environmental assessment (EA).

The EIS is the heart of NEPA. Under Section 102(2) (C) of the act, federal agencies must prepare a detailed statement, known as an EIS, for major federal actions significantly affecting the quality of the natural environment. The EIS must include a statement of: the environmental impact of the proposed action; any adverse environmental effects that cannot be avoided should the proposal be implemented; and alternatives to the proposed action.

EISs AND EAs

In practice, the EIS involves extensive environmental analysis, evaluation of all reasonable and practicable alternatives to the proposed project, considerable interagency review, takes 2-3 years to complete at considerable expense, and is normally several inches thick. In one extreme case, an EIS consisted of 24 volumes, each 300-500 hundred pages in length. Although EISs are usually required only for major projects, they are triggered not just by the size of the project but also by the value of the resources affected and the magnitude of the controversy.

An EA, however, is sufficient for most projects that fall under NEPA. It is usually a brief document of 5-10 pages that can be completed in a few days or weeks. Like a mini-EIS, an EA briefly describes the purpose of the project and the likely environmental impacts, offers an analysis of alternatives, and indicates whether or not the impacts will be significant. In most cases, an assessment results in what is called a finding of no significant impact (Fonsi), but in a few cases the impacts will be significant enough to warrant that an EIS must be prepared.

The EIS process can either harm or help drilling plans. In some cases, the process can delay project approval, cost hundreds of thousands of dollars, and consume the equivalent of a small forest of paper. Or worse, after all the time and expense of preparing an EIS, a project proposal may be rejected. For highly controversial projects, the EIS can become the lightning rod for opposition. It will be discredited if it fails to anticipate and address controversial impacts and reasonable alternatives, or if it was evidently conceived merely to justify the project.

On the other hand, the EIS process can help local, state, and federal agencies focus on environmental issues and cooperate in the analysis of real problems. In many cases it has helped project proponents channel criticism into constructive paths and useful studies. It has also improved decision-making efficiency by establishing firm schedules for organizing and analyzing data necessary under many different statutes.

The Council on Environmental Quality's regulations specify what impacts a federal agency should analyze in an EIS, such as direct and indirect effects of a proposed action; possible conflicts between federal, regional, state, and local land use plans, policies and controls; and the environmental effects of alternatives to the proposed action.

A federal agency may, however, have difficulty determining the appropriate scope of its NEPA analysis. For example, if a contractor proposes to build a large production facility on uplands and a service road through several acres of adjacent wetlands, he will need a 404 permit from the Corps of Engineers. Should the Corps of Engineers, in complying with NEPA, evaluate the environmental impacts of both the upland production facility and its service road through wetlands or only the road? Should it evaluate whether the adverse impacts could be reduced by constructing the entire production facility at a different location?

Recently the Corps of Engineers' NEPA analysis was significantly limited. Until 1988, the Corps of Engineers' NEPA policy has been to analyze the direct and indirect impacts of a project where a federal permit was required.

In 1984, the Corps of Engineers proposed to change its NEPA procedures in order to avoid situations in which non-federal issues were being implemented by the federal government. The new procedures would confine its analysis of a project's environmental impacts to that part of a project requiring the Corps of Engineers permit. In the above example, for instance, since a 404 permit was required for the road but not for the production facility, the Corps of Engineers would limit its NEPA analysis to the impacts of the road.

EPA objected to the Corps of Engineer's proposed changes in 1985 and referred the matter to the Council on Environmental Quality, which is supposed to resolve such interagency differences. In 1987, the Council on Environmental Quality finally settled the long-standing dispute and largely accepted the Corps of Engineers' proposed changes.

In February 1988, the Corps of Engineers issued rules to clarify and streamline its procedures for assessing the environmental impacts of projects under NEPA. The new rules significantly narrow the Corps of Engineers' scope of analysis and limit the reach of federal interest in the environmental impacts of private projects, including drilling projects in wetlands.

Section 404 permit applications are subject to the environmental impact provisions of NEPA, but the Corps of Engineers has estimated that less than 0. 5% of applications cover projects that because of their likely impacts on the environment, will require an EIS.

A wellsite that has been determined not to be located in an environmentally sensitive area must still be constructed with the idea of minimizing adverse environmental impacts. Although the regulatory requirements of drilling a well in a non-wetlands area are not as rigorous as for a wetlands site, the contractor should pay attention to the wants and needs of the local and state agencies and the landowner to ensure that the drilling project can be a success for all parties involved.

MITIGATION

Most problems that occur between the drilling contractor or operating company and the land owner(s) are caused by miscommunication as to the condition of the site before and after the well has been drilled. This miscommunication can add tens of thousands if not hundreds of thousands of dollars to a project. Often the sentimental value of the land comes into play, and no amount of money can offset any changes that are caused by the drilling operation. If the value of the potential damage is not negotiated up front, then burden of proof as to the condition of the land is on the contractor or operator. This is a costly way to do business.

This is why it is important for any contractor or operator to negotiate a settlement up front for damages that a drilling operation may cause. This is called mitigation. Mitigation can take many forms. It can be monetary, or it can be as little as keeping the grass mowed in an easement to the well site. A lot of this will depend upon the land owners' past experiences with wells on their properties and how much a contractor or operator is willing to give for a certain type of damage.

It is important for any contractor or operator not to set precedents that will make it difficult for future dealings in efforts to drill a well in that area. Do not expect to pay "Joe Landowner" $2,000/ac for damages on your first well and pay "Jane Landowner" $50/ac for damages on the second well in a particular area. Most land owners in an area share this information, and once a precedent has been set, it is hard to renegotiate settlements downward.

The mitigation process becomes more difficult when the landowner does not have the mineral rights for a piece of land. The attitude changes substantially when the landowner realizes that the only compensation that he is going to receive is a land damage settlement. The price of the property tends to increase almost overnight. On the flipside, landowners with mineral rights tend to invite you in for a meal or take you hunting and in some instances even offer to help clear the land for the drill site. Therefore it is important that prior to commencing the mitigation process with a landowner, that the company representative be aware of just what situation the landowner is in.

If at all possible, the contractor or operator should avoid mitigating with money. Once monetary mitigation in a trend area has been set, it is next to impossible to reverse it. Here are some ways that a contractor can mitigate with a landowner in non-wetlands areas:

- The first thing to do is to offer to bring the land to as near pre-drilling condition as possible. In some cases the landowner is willing to live with the inconvenience of a drilling operation as long as when the operation is complete, the land will be restored to as near original condition as possible. This is an inexpensive means of mitigating damage and something that a contractor or operator should be doing, anyway.
- Offer to do some light maintenance work on their land. Works such as grass mowing or fence fixing can go a long way with an eighty-five year old farmer who cannot do it himself.
- Do some terracing or landscaping of the land. This is a popular alternative in hilly areas. Since you will probably have a bull dozer on the location anyway, spend a couple of extra hours terracing a piece of land. This can save the landowner from doing it and in most cases, increase the value of the land.
- Put in a gate or cattle guard. On most pieces of property, especially if it belongs to a farmer or rancher, it is important to keep livestock in a certain area. Offering to erect a gate or place a cattle guard is a perfect solution for this problem. In most cases, you will probably have to erect a gate anyway for well site security reasons.

- Set a culvert(s). One of the single biggest problems in rural areas on land is drainage. If you need to cross a stream or drainage canal to get to your location, set a quality culvert and leave it in place when you leave. Used railroad train cars are good for this.

This is not to say that by utilizing these mitigation techniques you will be successful in keeping a landowner happy or that you will not have to make some type of monetary restitution for damages. This is simply saying to attempt these mitigation techniques first, before settling with money.

In wetland areas mitigation can become more difficult. Much of wetland acreage is controlled, governed, or owned by a federal or state agency. Therefore mitigation for the property that is being impacted may not be conducted on that property but rather on an adjacent property. This is a direct reflect of the "zero net wetlands loss" regulations that were implemented in the 1980s.

A common type of mitigation requested by the Corps of Engineers in wetland areas involves placing earthen plugs or dams at the entrance of old drilling slips that were cut for the purpose of drilling other wells in the area. The earthen dams tend to serve two purposes. First they prevent recreationalists, such as fishermen, trappers, etc., from entering the slips, and second, they reduce the amount of erosion that occurs in the slip area as a result of boats speeding by the slip. With the high rate of erosion that occurs naturally, any elimination of man-made erosion helps.

Other types of mitigation that are requested by the land owners of wetland acreage is the formation of artificial reefs by knocking down the ring levees at the drill site and allowing the site to flood. The drilling pad, which is usually topped with limestone, provides a firm bottom for the reef, which subsequently provides a habitat for fish and other wetland wildlife.

Most of the mitigation in wetland areas is more dictation than voluntary mitigation. That is, if you want to utilize the property for drilling purposes, you basically must abide by the wetland owner's wishes.

There is one way, however, that a contractor or operator can minimize the amount of mitigation required. That involves utilizing directional drilling techniques to drill a well from an existing drillsite location. This method pleases the land owner because no damage is being done

to his property; it pleases environmentalists because no new wetlands are being impacted; and it pleases the contractor or operator because no additional money for mitigation is required. That does not mean that this method is not without cost, however. Drilling a well directionally adds cost to the drilling of the well by increasing the total drilling depth, which in turn increases material costs and risk. With the increased drilling depth, the well will require more casing in the directional portion of the well. The risk of the drill string getting stuck in the well because of sidewall contact of the drill string also increases substantially. The contractor or operator should weigh the cost of mitigation against the cost of drilling the well directionally. After analyzing these things along with the impact the site will have on the local ecology, an informed decision should be made.

7

DRILLING TECHNIQUES

When a geologist, engineer, or layman, through maps, plats, logs, or just a hunch believes that oil or gas may exist in a certain area, the last thing on his or her mind is where the surface location is. Outside of offshore locations, which is another matter, the determination of whether or not the well can be drilled on the most advantageous and most cost-effective surface location can be a challenge.

There are several factors that govern where a well can be drilled to take advantage of natural geologic trapping mechanisms and the surface location. For instance, if a desired geologic location at 15,000 ft is situated directly beneath a residential neighborhood, the likelihood is remote that the contractor will be able to locate the drilling site directly over the geologic target in the front yard of someone's house.

Similarly, if a desired geologic location is situated in the middle of a lake or a swamp and utilizing a drilling barge to drill the well doubles the drilling costs, it would behoove the contractor to compare the cost of utilizing directional drilling techniques to drill the well to its objective or at least compare the cost of directional drilling with that of the drilling barge technique.

In most cases, after weighing the factors, for successfully drilling a well utilizing directional drilling techniques against those for a straight hole, the ultimate deciding factor will probably be the cost.

Some reasons for drilling a well from a particular location are mandated by government agencies with jurisdiction over the environmental aspects of an area. This regulatory involvement has been growing with the increased concern over the effects of oil and gas drilling practices on environmentally sensitive areas. Wetland areas have drawn the biggest attention in recent years. Since the passage of wetlands regulations concerning permitting and usage, the determination of a site as being environmentally sensitive has been of major concern to drilling contractors and well operators. The determination of a site as being located in an environmentally sensitive area can cost the contractor or operator hundreds of thousands of dollars in incremental drilling costs for utilizing techniques such as directional drilling or closed loop systems that will help minimize the impact.

New Approaches

When a geologic target is situated in such a way that the surface location cannot be placed directly above it, whether for its proximity to residential areas or because of the Army Corps of Engineers or the land owner does not want any further dredging to occur in a certain area — directional drilling techniques will need to be employed.

In a literal sense, wells have been drilled directionally for ages. No well has ever been drilled perfectly straight, even when it was intended to be drilled straight. When drilling a well, the natural physical laws tend to overtake the intent. Drill strings, being extremely flexible and maneuverable, tend to follow the natural dips in the formations that are being drilled. If not monitored through check shot surveying at controlled intervals, a bottom hole location can end up several hundred feet if not thousands of feet displaced from the surface location with bore hole angles of 10-60°. When a company is drilling for a geologic marker or trap that is only a few hundred feet in length, not knowing the well bore's direction can miss the objective entirely.

Directional drilling techniques have improved tremendously in the past 15 years. Modern directional drilling instruments can indicate what angle and direction a bore hole is located and even predict, based on current drilling parameters and formation characteristics, which way the drill string will tend to go.

These modern directional drilling techniques have made it possible for a surface location to be located 1-2 mi from the subsurface geologic target and come within inches (based on the accuracy of the instrumentation used and the input data) of the desired bottom hole location. Furthermore, with the combination of downhole drilling motors, measurement while drilling instrumentation, and directional instrumentation, wells can be drilled at extremely high angles and geologically logged while being drilled all in a single pass.

In recent years, the concept of drilling wells horizontally has undergone widespread acceptance. A horizontal well, which is drilled at an angle of 90° or greater, can successfully expose more of the targeted geologic formation to the well bore, thus allowing a much greater oil and gas flow rate. While horizontal and other extended reach wells cost more than vertical wells, the several-fold increase in production more than justifies the added cost.

These modern drilling methods have made it possible for a company to drill wells to geologic objectives that once were bypassed due to the lack of access to a suitable surface location because of environmental concerns.

All too often a drilling contractor or operator will get into a mindset that "bigger is better:" a bigger bit, a bigger rig, and a bigger location. In the "good old days," this was not only the way to be successful in drilling wells but also an acceptable philosophy with the government agencies.

As environmental laws have changed, so has that philosophy. This philosophy changed to "leaner and meaner" — or perhaps, "less is more." What this meant to the contractor or operator is that in areas deemed to be environmentally sensitive, drilling locations were going to be limited and the utilization of massive drilling rigs to drill shallow wells would have to change.

Some companies heeded the call and began to search for other ways to accomplish the same goal with only half the rig. Instead of a rig that required a 300 ft by 300 ft location, with new environmental regulations, the contractor was going to have to find a rig with the capability of drilling on a drill pad half that size.

This was accomplished not by reinventing the drilling rigs but by utilizing existing drilling rig technology. For many years, drilling companies maintained a fleet of mobile, truck-mounted drilling units. For the most part, they were utilized for the drilling of extremely shallow wells or for working over wells. When the new environmental rules were implemented, contractors began to test the drilling limitations of these large workover units. By reviewing the mechanical capabilities of these units and modifying them to withstand slightly heavier loads that are a result of deeper wells, contractors began to utilize these drilling units to drill deeper wells in environmentally sensitive areas.

These drilling units gave companies the flexibility to drill from locations that are only half the size of the pre-environmental push drilling locations and at the same time save 30-50% of the cost of drilling these wells. The main rig components of the smaller drilling rig that must be reviewed to ensure that it will be capable of drilling a deeper well are its derrick capacity, substructure capacity, rotary table capacity, mud pumps and drilling fluids system.

Operating Concerns

When a well is drilled below 12,000 ft, it is more than likely that a larger grade of drill pipe will be required and that multiple strings of casing (rather than just surface casing and production casing) will need to be run because of higher geopressures.

Deeper wells create higher friction pressures and higher torques that are subjected to the drill pipe. When drilling shallow wells, 3 in. or 4 in. drill pipe usually is adequate to drill to total depth. With deeper wells, the higher torques require that larger drill pipe be used. The larger drill pipe will provide the contractor with adequate strength to avoid pipe damage such as twisting off of the drill pins.

The formation geopressures that are encountered on deeper wells usually require that multiple strings of casing be run to protect shallower zones from overexposure and to allow adequate fracture pressure at the casing shoe to prevent lost circulation and a possible underground blowout.

Both the larger drill pipe and the additional casing have one thing in common: the derrick must have the capacity to lift and run them into the well. While drilling, a margin of overpull to overcome friction pressure and potential stuck pipe operations will be required. On larger rigs, the limitation for overpull is usually in the drill string or casing. In the case of a smaller rig, one with a smaller derrick, the limitation is shifted to the derrick. A typical derrick maintains a 100% safety factor for its capacity limitation. As a rule of thumb, a contractor should allow at least 100,000 lb of overpull on any anticipated string of drill pipe or casing.

If through the review it is determined that the derrick of the particular truck-mounted drilling unit does not maintain an adequate amount of bearing capacity, modifications to the derrick, such as bracing or shoring, can be made. But ultimately if the company wants to ensure that the derrick maintains the proper bearing capacity to handle the drilling and casing running loads that it will be subjected to, it may have no choice but to replace the derrick. This may be a costly alternative but probably will be more economic than buying a new rig.

If the derrick capacity is adequate to sustain the drilling loads but is not adequate to sustain the casing running loads, the contractor has two alternatives that may allow him to not have to modify the rig. The contractor can float the casing in the borehole while running it or he can utilize casing jacks to run the casing.

In a typical casing run operation, the contractor fills the inside of the casing with fluid as is it is being run. This is usually required because a cementing shoe with a one-way check valve is run at the end of the casing string. This serves two purposes: first, it allows the casing to be run faster than if the inside of the casing is empty; second, if a pressure or salt water kick is taken during the casing running job and the one-way check valve fails, the full head of drilling fluid will help hold back the influx of gas or fluid from the geopressured zone. Running the casing without filling it inside, or "floating," with drilling fluid allows the buoyant forces being subjected on the casing to lighten the weight of the casing string, thus allowing the contractor a larger margin of overpull at the surface and on the derrick.

If the contractor is uncomfortable with floating the casing in the well, he has the alternative of utilizing casing jacks to run the casing. The casing jacks are independent of the rig's substructure or derrick and can be designed to handle the casing independent of the rig or in combination with the rig. Although cumbersome and slow, the utilization of casing jacks can prevent unnecessary damage to drilling unit components and prevent the contractor from making expensive permanent modifications to the drilling unit.

Like the derrick, the substructure is a critical load bearing component of the drilling unit. The same loads that are imposed on the derrick from drilling deeper wells are imposed on the substructure. If the contractor is faced with a possible inadequate load rating on the substructure, the same alternatives as for the modifying of the derrick are available. The contractor can modify the substructure, change out the substructure, float the casing in the hole, or utilize casing jacks to run the casing.

When drilling deep wells, a tremendous amount of friction is being generated on and by the drill string. This friction is felt at the surface by the rotary table as torque. The purpose of the rotary table is to provide enough torque at the surface to rotate the drill string and continue drilling.

Rotary tables come in a variety of sizes, depending upon the gear ratios, the driver, and the needs of the drilling unit. A typical large rotary table has the capability of providing as much as and in excess of 30,000 ft lb of torque to the drill string.

In many instances, when a deep directionally drilled well is nearing total depth or if severe dog legs are present in the wellbore path, the torque that is required to drill the well exceeds the capabilities of the rotary table. When this occurs, the contractor is left with either changing out the rotary table for one with higher torque generating capacity, utilizing a downhole motor to rotate the drill bit, or manipulating the drill string configuration to help minimize the friction in the drill string.

Changing out the rotary table in the middle of a drilling operation can be time consuming and detrimental to the open hole. It typically takes 12 hr or more for the rotary table to be changed out during drilling operations.

Because the rotary table is the center of the well bore and is utilized to hold the drill string in the well, all of the drill pipe must be removed from the well to conduct this operation. This means that the borehole will be left open during the entire rotary table exchange or repair.

There are basically two risks of leaving the borehole open without circulating fluid in it. The success of most drilling operations depends upon the ability of the contractor to clean the cuttings out of the well and avoid bore hole loading. When the drill string is out of the well, no fluid is being circulated in the well thus allowing a bigger change for cuttings to build up in the bore hole.

This, in combination with the angles that are associated with the well path, can effectively close up the bore hole. If this occurs, after the rotary table has been repaired, redrilling or reaming may be required to remove the bridge(s) that are blocking the well bore path. Depending upon the severity of the hold blockage, it could take several days to regain access to the well bore. In well bores where the formation clays are time sensitive due to overexposure to the drilling fluids, formation swelling may force the drilling contractor to conduct a tremendous amount of reaming to open the well bore to its original size and gain access to the bottom of the well.

Because the rotary table's limitations are met usually when the well is near its objective, the possibility of a kick occurring substantially increases. One of the keys to successfully subduing a kick is to have the drill pipe at the bottom of the well to enable the circulation of kill weight fluid around the whole well bore.

When the drill string is out of the hole for a long period of time to repair the rotary table, an appreciable amount of gas could build up in the well bore and induce a kick. Without having the drill string at the bottom of the well bore to circulate kill weight fluid into the well bore, the drilling contractor loses the ability to effectively battle the kick. If a kick does occur at this point, the drilling contractor is left to utilize the live well lubricating or snubbing techniques to get the drill string back to the bottom of the well. A snubbing operation can take several days to complete and is not guaranteed to be successful.

Knowing the potential problems that can occur by having to change out a rotary table during a drilling operation, the drilling contractor should make sure that the torques anticipated during the drilling operation have been explored and that the capabilities of the rotary table have been reviewed. If this review indicates that torques exceeding the limitations of the existing rotary table exist, either the rotary table should be changed out before the well is drilled or a different drilling rig should be considered for this job.

Some drilling contractors choose not to change out the rotary table when its limitations are met but opt to utilize downhole drilling motors to turn the drill bit. A drilling motor is basically a typical rotor and stator motor that is operated by the passage of drilling fluid through it. Many configurations of the rotor and stator can be used to meet the downhole torque needs of the drilling operation. By increasing the number of rotor and stator vanes, the drilling contractor can increase the torque capacity of the downhole motor. The drilling contractor can increase the torque capacity of the downhole motor. The drilling contractor also has the ability to chose either a high speed-low torque motor, for shallower, easier drilling or a low speed-high torque motor for deeper, more difficult drilling.

The motor is placed at the end of the drill string, and the drill bit is placed at the end of the motor. Once the mud pumps are turned on, the drilling fluid passes through the motor, through the rotor and stator and turning the drill bit. The drill string does not have to rotate, thus not requiring the rotary table to be turned. Once the objective has been reached, the motor can be pulled and the well logged and completed.

Instead of changing out the rotary table or utilizing downhole drilling motor or even using a combination of these, the drilling contractor often chooses to reconfigure the drill string to overcome torque problems. This can be done in a couple of ways. A tapered drill string, with the smaller drill pipe downhole, can be run; or aluminum drill pipe, which has only half the weight of normal drill pipe, can be run.

A tapered drill string allows the drilling contractor to lessen the weight of the drill string and thus reduce the friction that is being generated in the well bore. On a typical 15,000 ft highly directional well, a tapered drill string could consist of the bottom hole assembly, 5,000 ft of 3½ in. drill pipe, and 10,000 ft of 5 in. drill pipe to surface. The 5 in. drill pipe is run at the surface because of its higher torque rating. The only inconvenience that is experienced utilizing a tapered drill string is that a separate set of drill pipe handling tools will be required to handle the 3½ in. drill pipe versus that for the 5 in. drill pipe.

Basically the aluminum drill pipe, whose weight is less than half the weight of conventional steel drill pipe, is run at the bottom of the drill string. As with the tapered drill string on a typical 15,000 ft well about 5,000 ft of aluminum drill pipe will be run on bottom with the remaining 10,000 ft consisting of steel drill pipe for torque requirements.

Many drilling contractors have experienced as much as a 60% reduction in torque by utilizing aluminum drill pipe in high angle, high torque wells. As with the tapered string, special drill pipe handling tools are required to run the aluminum drill pipe. In addition, because of potential unwanted chemical reactions between the drilling fluid and the aluminum drill pipe, the pH of the drilling fluid must be kept between 6 and 8 to avoid the aluminum drill pipe from pitting.

DRILLING WASTES

All industries regardless of their functions generate some form of waste. The physical theory that matter is neither created or destroyed, it just changes forms is a perfect analogy for the creation of waste. That is to say, the creation or use of any product also creates waste. Even if the end product is a clean burning fuel, somewhere along the line byproducts of its process were created. Since hydrocarbon products have a wide spectrum of uses because of molecular makeup, distillation, fractionation, and other forms of processing create a tremendous amount of less desirable or lower grade byproducts. What to do with or how to dispose of these byproducts becomes a major issue.

With more developing countries getting involved in the exploration for and exploitation of oil and gas reserves, the issue of how to handle this waste is growing. At the same time, countries such as the United States that have been exploiting oil and gas for many decades are now taking measures to help reduce or minimize the amount of waste that is created by these operations.

This poses an extremely difficult dilemma. Should these developing countries be burdened with the expense of complying with more stringent, costly drilling waste minimization methods when during the early years of the U.S. oil industry oil and gas was being produced essentially with minimal awareness of environmental impacts? Such concerns are being addressed at the intergovernmental level as well as by international lending institutions and aid/development agencies.

This growing awareness of social responsibility and public interest has caused waste generation, handling, and disposal to command a higher priority in government regulations. Although the oil and gas industry has enjoyed some exemptions from the more stringent waste regulations, other regulations have cost the industry hundreds of millions of dollars.

With this increased scrutiny, almost all U.S. oil industry waste generators have become extremely aware of the need to implement waste minimization plans in all of their operations that have the potential to create wastes.

It is evident that no oil or gas well can be drilled or produced with zero waste. The installation of special equipment, the introduction of special techniques, and most important, the implementation of training programs to minimize waste can have a profound influence on the amount of drilling waste that is generated at a well site.

Equipment such as closed-loop drilling systems and recycling systems have been successful in reducing the amount of waste that is generated at well sites by as much as 50%.

Special techniques such as chemical enhancement of the drilling fluids or the use of smaller, more efficient drilling rigs has minimized the impact a drilling site has on the surrounding environment.

Many companies realize that the key to successfully implementing any waste minimization program is through the education of its work force. Onsite as well as offsite training should be given to almost every employee that works on a rig to enable him or her how to recognize as well as handle waste, thus minimizing its generation. Many companies have commenced incentive programs that reward those employees who develop and implement successful waste minimization programs.

Some oil and gas industry personnel, however, feel that this increased scrutiny of waste minimization has placed an unnecessarily large financial burden on the industry. Training and re-education of personnel can be costly and implementation of these plans even more costly.

Many domestic oil and gas drillers and producers have ceased to drill in the United States because they contend the extremely stringent domestic environmental policies are too costly to make their drilling ventures profitable. Many of these drillers and producers are abandoning domestic operations to drill and produce oil and gas outside the U.S. They feel that the less stringent regulatory oversight elsewhere allows them to be competitive in today's oil industry when oil and gas prices no longer command a premium.

There are ways, however, that companies can modify their current operations to meet the requirements of these stringent environmental laws as well as meet their financial goals. The key to their success in achieving environmental compliance, in most cases, is by merely becoming more efficient in their operations.

Those operators and contractors not actively pursing avenues to become more efficient in their operations are having to pay a higher price for the handling and disposal of excessive wastes generated as a result of these inefficiencies. For the most part, the implementation of programs to minimize wastes is fairly inexpensive.

In most cases, the only program that needs to be implemented is that of a change in philosophy of how operations should be conducted. This can be done through on-the-job training, seminars, and training classes. Even for the most expensive waste minimization plans, the costs to implement them is more than offset by the drastic reduction in waste that must be hauled off to an expensive disposal site or the utilization of an expensive remediation technique.

There are basically two categories that oil field wastes are listed under in the Resource Conservation and Recovery Act (RCRA): non-hazardous and hazardous. Unless unauthorized chemicals and solvents are being utilized during the drilling of a well, usually at least 95% of all wastes generated at a drill site can be classified as non-hazardous oil field waste. The remaining 5% is classified as hazardous waste.

For anyone that has worked in the oil and gas industry, the differentiation between non-hazardous oil field waste and hazardous oil field waste may seem intuitively obvious. But then again we are talking about an industry whose public image is filtered through media images of oil soaked birds and choking baby seals.

In reality, though, it is an industry that spends millions of dollars on equipment and training to prevent environmental disasters from occurring.

The biggest hurdle the industry faces is that of educating the general public and the government as to these efforts and the differences between non-hazardous oil field wastes and hazardous oil field wastes. This lack of understanding has caused many state and federal laws to be passed that prohibit the drilling of oil and gas wells within their jurisdiction as well as anti-oil activist groups to protest the drilling of oil and gas wells in certain areas. This opposition often causes companies to spend unnecessary money to persuade or fight this opposition. In some areas, this can lead to a company spending additional hundreds of thousands of dollars attempting to mitigate these differences. This often can cripple a drilling project to the point of making it an unprofitable venture. This is why it is important to make clear the differentiation between non-hazardous and hazardous oil field wastes clear.

Non-Hazardous Oil Field Waste

The legal definition of non-hazardous oil field waste(s) as recognized by most government environmental agencies is: any garbage, refuse, sludge from a waste treatment plant, water supply treatment plant, or air pollution control facility; other discarded material, including solid, liquid, semi-solid or contained gaseous material that does not cause or significantly contribute to an increase in mortality or an increase in serious irreversible or incapacitating reversible illness or does not pose a substantial present or potential hazard to human health or to the environment when improperly treated, stored, transported, disposed of, or otherwise managed is considered a non-hazardous waste.

Although any definition of non-hazardous oil field waste is very explicit as to how a waste is determined to be non-hazardous, there are times that the classification of a waste as hazardous or non-hazardous is left to be subjectively decided by a government official that has never seen a drilling operation or has no knowledge of the industry. This makes it extremely important for company personnel (onsite foreman, driller, toolpusher, engineers, and rig workers) to understand exactly what type of waste is being generated before it is mixed and disposed of. Inadvertent misclassification of a waste generated on a drilling location can cost a drilling company hundreds of thousands of dollars in hazardous waste disposal fees. The laws governing hazardous wastes usually do not allow for the lack of knowledge by an individual or company as to the classification of a waste. This lack of knowledge can lead to exorbitant fines and criminal penalties.

Fortunately the oil and gas industry currently enjoys several exemptions for several types of wastes generated in operations from hazardous waste classification. This is done so under RCRA. Under RCRA, there are several acceptable methods for disposal of standard oil field wastes, including drilling fluids, drill cuttings, and produced water. If applied properly, these methods are fairly inexpensive and require very little high tech equipment to implement.

Another reason for having a thorough understanding of what types of wastes are being generated at a drill site as well as attempting to minimize the amount of waste that is being generated is that proposed RCRA legislation seeks to reclassify most oil field wastes as hazardous waste. A change in the classification of wastes would be economically devastating to the industry.

There are many sources of non-hazardous wastes in an oil or gas field. A drilling operation with a drilling rig, engines, equipment, and living quarters is much like a small city generating various types of wastes. This totally self-contained drilling operation generates waste much in the same manner that any individual neighborhood does. Waste such as paper and plastic wrappers, bags, domestic waste fluids and sanitary wastes are common to any drilling operation. Other types of wastes that are generated on a drilling operation are drilling fluids, drill cuttings and used oil. Like any wastes, if not handled and disposed of properly, their impact to the environment can be significant.

Drilling Fluids

A drilling fluid has many uses in drilling a well. A drilling fluid is used to transport drilled cuttings out of the bore hole, to cool the drill bit to prevent bit failure caused by heat, and to hold back formation geopressures. Utilizing a fluid to drill with is a concept that dates back to ancient times. These early drillers used water as their drilling fluid. Although these wells were fairly shallow, the use of a drilling fluid was just as important then as it is now.

Drilling fluids used in these early wells were nothing more than fresh or salt laden water. Water was abundant and free and was thought of as being non-damaging to the bore hole and the environment. Although these fluids did not possess the best properties for cuttings transport or formation inhibition, they did provide a means for the driller to circulate the bore hole, cool the bit, and transport the cuttings.

The drilling fluid also was easily disposed of. Since water is a natural substance, it was thought that the drilling fluid could be disposed of anywhere. Often the fluid was discharged to the ground or into a surrounding body of water. This can be thought of as the first form of land farming.

The concept of land farming entails allowing drill cuttings and returned drilling fluid to be spread over a section of land and tilled until the concentrations are reduced to non-damaging levels. In modern times, land farming (which will be discussed in detail later in this book) is utilized with shallow wells or in the shallow portion of wells to dispose of the drilling fluids and/or cuttings. This technique is usually used in boreholes as deep as 8,000 ft. Beyond this depth, environmentally damaging chemicals such as barite, lignosulfonate, and other chemicals often are added to the drilling mud to attain certain drilling fluid properties. In most cases, after a farmer finds out that the drilling fluid acts as a fertilizer, there often follows a request that the drilling fluid that passes all criteria be land farmed on his property.

The utilization of plain water as a drilling fluid does have its limitations, however. In most areas where wells are drilled, exposing the formation to fresh water can cause the exposed clay (e.g., silica and alumina) to swell and become unstable. This swelling can cause many other drilling problems such as stuck drill pipe and lost drilling fluid returns.

On many offshore wells where the formations are geologically young and active, severe swelling of the clays can occur, thus causing the infamous "gumbo attacks." These attacks are caused by the fresh water coming into contact with the kaolinite, illite, cholrite and montmorilonite clays in the bore hole. During the gumbo attacks, the hydrophilic clay cuttings can swell several times their original size. This swelling has been known to cause the drill string to become stuck, cause the loss of drilling fluid returns, plug the return line from the bore hole, and push the rotary bushings out of the rotary table. In all of these instances, a lot of time and money can be spent trying to combat these problems. Unplugging a flow line or freeing stuck pipe can be extremely expensive on a well. Special equipment such as gumbo slides, high pressure water sprayers, and racks have been developed to try to minimize the effects of the gumbo.

To minimize the clay swelling problem, contractors have utilized several methods. Two straightforward methods involve adding caustic soda (sodium hydroxide, NaOH) or adding sodium chloride or potassium chloride to the drilling fluid.

The addition of caustic soda to the drilling fluid causes the pH of the drilling fluid to become alkaline or "hot. " This technique dissolves a majority of the clays (gumbo), leaving the dissolved clay in solution. Although it works great in minimizing a gumbo sticking problem, it leaves an extremely high level of fines (colloidal size clays) in the drilling fluid. This can substantially increase the chemical treatment cost of the drilling fluid later in the drilling of the well.

The addition of sodium chloride or potassium chloride to a drilling fluid increases the salinity of the drilling fluid to at least that of the natural formation that is being drilled. The fluid will then inhibit the clay and prevent the clays in the formation from swelling. The problem this poses is that now that the return fluid maintains high salinity values, so the return fluid and cuttings now must be disposed of at an approved landfill or by another approved method and cannot be land farmed. Being that thousands of barrels of fluid are generated in this process, the cost to dispose of the fluid can be substantial.

As a well is drilled deeper, another factor, such as increased formation pressure, becomes an issue. A driller's first line of defense in containing these pressures and not allowing an influx of formation fluids into the wellbore and possibly causing a blowout is that of proper drilling fluid density. Fresh water has a fluid density of approximately 8. 34 lb/gal. This density is adequate for shallow wells or in the shallow portion of deep wells. But in deeper wells or in areas where shallow abnormal geopressures are encountered, heavier drilling fluid weights are required to contain formation fluids. Typically in areas where abnormal geopressures are encountered, drilling fluid weights of 12-20 lb/gal. are required to contain the formation fluids.

There are several ways that drilling fluid density can be increased to meet the proper overburden pressure. The two most popularly utilized methods of increasing drilling fluid density are through the use of drill cuttings and the addition of weighting materials to the fluid.

Through bypassing a certain amount of the return drilling fluid from the drilling rig's solids control system, the entire drilling fluid system can be increased to about 10 lb/gal. This is a popular method of increasing drilling fluid density when the background gas or a slight increase in fluid influx occurs in the shallow portion of a well.

Although the drilling fluid properties are not ideal when this technique is utilized, it allows the drilling contractor to save a substantial amount of money on the chemical costs for this portion of the well. There are drawbacks to this method, however. If the driller intends to utilize this drilling fluid in drilling the deeper portion of the well, or if well problems require a longer amount of time for a particular portion of the well, an extremely high percentage of solids can accumulate in the system. This can lead to bore hole loading. Bore hole loading can lead to other well problems such as drill string sticking and drilling fluid maintenance problems. In a drilling fluid system with a drill cuttings content greater than 15%, chemical costs can be as high as 100% of a drilling fluid with a less than a 15% drill cuttings content. This increased drill cuttings content also can cause problems in the drilling fluid pumps, shakers, and other solids control equipment utilized to maintain the properties of the drilling fluid.

The addition of weighting materials to a drilling fluid in order to increase fluid density is the most effective method. Carefully controlled material properties and years of research allow the proper use of these weighting materials to be used on even the most difficult wells.

Common weighting materials used in the petroleum industry are barite (barium sulfate) which has a specific gravity of 4.2 (that is 4.2 times the weight of water); siderite (iron carbonate), which has a specific gravity of 3.8; iron phosphate, which has a specific gravity of 5.2-6.4; galena (lead sulfide), which has a specific gravity of 7.0; hematite (iron oxide), which has a specific gravity of 5.0; and ilmenite (iron titanium oxide), which has a specific gravity of 4.6.

Chemically, none of these minerals maintains an electrical charge and thus are considered inert.

Each product has a different application in drilling a well. Much of what drives the use of a certain weighting material over another is the cost of the material. For example, ilmenite can be as much as three to four times the cost of utilizing barite for weighting material. But the use of barite has its limitations when the well gets deeper and hotter, and the drilling fluid weight needs to be increased to more than 19 lb/gal.

Many of the wells drilled in the oil and gas industry are drilled with a drilling fluid weight of 12-18 lb/gal. Barite has proven to be an exceptional weighting material for this range. But like most products, barite has its limits. With wells in which temperatures are expected to exceed

300° F. and where drilling fluids are greater than 18 lb/gal. , barite begins to lose its effectiveness. The elevated temperatures or extremely heavy drilling fluid weights require that materials such as hematite or ilmenite be used to drill the well to total depth. These materials are stable to about 400° F. The problem that is encountered, however, is that these materials are extremely abrasive on the drilling equipment and can increase the maintenance cost of all of the equipment.

Another material that is commonly utilized when the well temperature is extremely high or if formation clay swelling is a big problem is oil-based drilling fluid. The use of oil-based drilling fluid in many wells enhances the drillability of the well. The drilling penetration rate is improved by the use of oil-based drilling fluid, and clay swelling is kept to a minimum.

Using oil-based drilling fluid does pose problems, however, in terms of the risk of hydrocarbon contamination and disposal of the oil-based fluid. Unlike water-based drilling fluids, this approach generally utilizes diesel as its base fluid. An operator or contractor basically must accept that all of the cuttings and drilling fluids now will have to be hauled off or disposed of down the annulus. Since the base fluid is a hydrocarbon, land farming of the drilling fluid and the cuttings has been eliminated as an option. Therefore, a drilling contractor or operator must weigh the options of increasing the drillability of a well to the cost of potentially hauling off of all of the drilling fluid if an oil- based drilling fluid is utilized.

There have been many years of research and development involving the various materials used in drilling fluids. The biggest problem has been not in selecting the proper material to increase the fluid density but in determining how to properly dispose of the fluids. This is a problem, because high concentrations of these materials can be extremely toxic.

Although each state has different standards, there are generally limits as to the density and composition of the drilling fluid that can be disposed of either by land farming or discharge into an adjacent body of water. Although presently there are no agencies governing oil and gas operations that require approval and proof of a site closure, they maintain the right at any time to inspect the site or the contractor's files to determine if the closed site meets the specifications outlined in the regulations. The agencies maintain the right of inspection indefinitely.

Selection of a site for inspection is usually done at random.

Therefore, it behooves a contractor to ensure that a site is thoroughly environmentally sound before it is permanently closed. If the agency inspects the site at some point in the future and contamination is found, the burden of proof of when the contamination occurred is on the contractor. If closure reporting maintained by the contractor indicates that the site was free of contamination after its operations were abandoned, then it is the agencies' responsibility to determine how and by whom the site was contaminated. If it is determined that the contractor did not remediate the site back to its original state and is responsible for the contamination, then the contractor will be responsible for remediating the site to its original state. If the contractor's records indicate that the site was free of contamination at the abandonment of the operations, and the agency investigations determine that there is contamination present, then more detailed investigation as to the actual cause of the contamination will be forthcoming.

Therefore the key in closing a site that has been a site that may or may not have been contaminated with drilling fluids, is that pre and post-sampling of the site do not exceed the strictest of local, state or federal governmental regulations.

Drill Cuttings

The next largest volume of waste (next to drilling fluid) in drilling a well is that of drill cuttings. On a typical 16,000 ft deep well, as much as 4,000 bbl (including washout) of cuttings can be generated. These drill cuttings are usually just combinations of sand, silt and clays and other naturally occurring minerals. Therefore, land farming or trenching and burying are usually the typical methods utilized to dispose of them.

As with drilling fluids, however, the deeper a well is drilled, the more difficult it is to properly dispose of the cuttings. In the case of drill cuttings, the deeper the well is drilled, the more weighting or specialty chemicals are used, thus contaminating the cuttings. No longer can the cuttings be disposed of just by land farming.

If drill cuttings are contaminated with either formation hydrocarbons or oil-based fluid hydrocarbons, it will be necessary to dispose of the cuttings in another manner. One method that is employed in the industry to dispose of contaminated cuttings involves slurrification of the cuttings and injection of them into the annulus of an abandoned well bore or the current well bore. This process allows the contractor to dispose of the drill cuttings without paying to have them hauled off.

Increased radioactivity of drill cuttings is common when drilling older formations. This poses a problem in handling of cuttings, which must then be classified under naturally occurring radioactive material (NORM) regulations. Most states in conjunction with the Nuclear Regulatory Commission have developed regulations that strictly govern the storage, disposal, and handling of NORM. Special training and handling techniques must be followed in dealing with NORM. This will be discussed in more detail later in this manuscript.

Washdown Fluids

Keeping a rig clean is one key to minimizing accidents and injuries. During a drilling operation, thousands of gallons of water are used to clean the rig floor and other components of the drilling rig.

Just about any time a connection is made or the pipe is being pulled out of the ground, barrels of drilling fluid are going to be spilled. Many drilling rigs are equipped with mud-saving devices that attempt to minimize the amount of drilling fluid that is wasted. Nonetheless, an appreciable amount of fluid is lost. Ideally the drilling fluid is diverted back into the drilling fluid system. But in reality, much of the drilling fluid along with the washdown water used to clean the rig floor ends up in the cellar or on the location. Most cellars are several feet deep, so the impact of contaminated washdown water on the location can be substantial.

Not only can the washdown water contain drilling fluids, it also can contain oil and grease from the lubricating materials used on the rig; lead, zinc, and other metals found in the drill pipe dope; and aromatic solvents such as trichlorethylene to help cut the oil and grease.

As was discussed, the drilling fluid contains weighting materials that are regulated by local, state, or federal agencies in certain concentrations. Its spillage, which is inevitable, usually creates "hot-spots" on the location.

On some locations, these wastes are pumped into the rig levee around the location or in a holding pit and allowed to separate. The water is then discharged to the land or water surrounding the location.

But in most cases, the water does not meet discharge standards set by local, state or federal regulations. Most states have specifications that must be met before wastewater can be discharged from a location. This program is called the National Pollutant Discharge Elimination System (Npdes). If the state in which the discharge is sought does not have Npdes authority, EPA maintains specification authority. These specifications include parameters such as biological oxygen demand (BOD), total suspended solids (TSS) and alkalinity (pH). Wastewater not meeting these criteria must be treated and tested prior to being discharged. If treatment is not possible or desired, the wastewater then must be hauled off to an approved disposal site.

One way to help minimize the impact of washdown water on a location is to place an impermeable liner beneath the rig. For a drilling location, it is usually recommended that at least a 30 mil liner be used. The 30 mil liner is pliable enough to be installed easily yet durable enough to withstand drilling rig activity. Although this does not guarantee that no washdown water will get on the ground, it does minimize the chance.

Another way to minimize the impact of washdown water is to keep the cellar pumped down. This minimizes the accumulation of weighting materials, oil, grease, and other contaminants in the cellar that cause hot spots. A special tank should be placed on the drilling location to hold the washdown water for either haul-off or annular injection into the well bore.

Sanitary Waste

Because it is not an obvious waste that is generated on a drilling site, the handling and disposal of sanitary waste is often overlooked. Nonetheless, the guidelines that must be adhered to by sanitary waste generators are governed by the Clean Water Act of 1977, and penalties for noncompliance of these specifications can be substantial.

Sanitary wastes on a drilling location come from various areas. On an offshore location, the living quarters is the main generator of sanitary waste. On a typical offshore platform or jack-up rig, there can be as many as 60-70 personnel on the drilling rig at one time. This can lead to a tremendous amount of sanitary waste that must be treated.

Regulations governing the discharge of sanitary waste into U.S. oceans, gulfs, or seas require that certain chemical characteristics be met prior to discharge. Typically total suspended solids (TSS) and residual chlorine are the parameters that must be controlled.

With the large amount of sanitary waste that must be treated on an offshore rig, a system that is capable of maintaining the correct effluent parameters is critical. The typical sanitary waste system that is utilized on offshore drilling rigs is a "red fox" unit (Figure 8-1). This unit is specially designed to handle large amounts of wastes while maintaining the proper effluent specifications. It maintains enough retention area and a series of weirs so that proper settling and segregation is accomplished. It also maintains a calibrated chlorine dispenser that allows the proper amount of chlorine to be maintained in the system at all times. Maintenance on the system is also easy. The system is designed so that chlorine concentrations can be maintained at the proper level to properly disinfect the waste and kill bacteria.

Upon initial installation, the system should be tested to ensure its proper operation and to ensure that the proper amount of chlorine has been added. A quick chlorine test can be run onsite, and necessary adjustments made to the system. Once it is determined that the system is functioning properly and the required effluent parameters are being met,

FIGURE 8-1

RED FOX UNIT

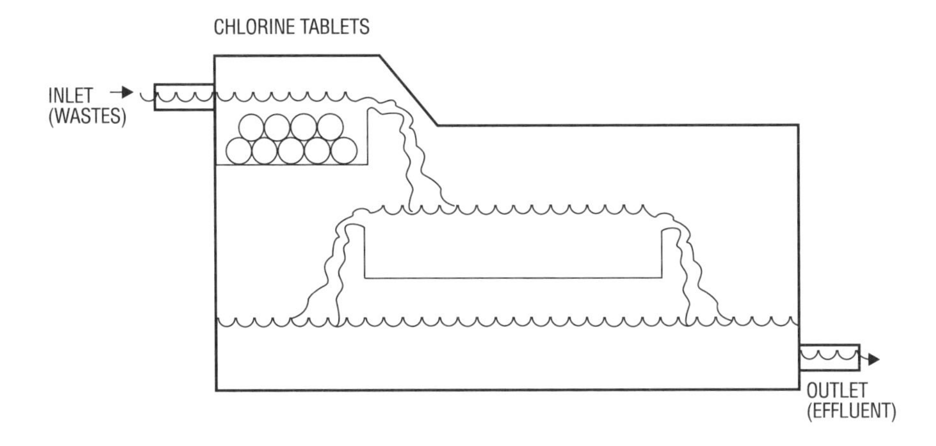

then the system can be operated with minimal maintenance.

On land locations, sanitary waste treatment and disposal can be a more sensitive issue because the impact of mishandled waste appears to be more tangible. The biggest reason is the accessibility of these wastes on a land location. On an offshore location, although not of any less importance, the amount of discharged waste relative to the receiving body of water is extremely small. And if the waste did reach an area that can be affected, the waste has been extremely diluted. On land, however, mishandled or mistreated waste directly effects either land, animals or humans.

Like the offshore rig, a large number of personnel can be working on a location at one time. This leads to a tremendous amount of waste generation. Because land drilling locations are usually more temporary than offshore platform locations — or at least the treatment facilities are — the systems used on land locations are usually more primitive than offshore systems.

A typical treatment system used on a land location is that of a honey pot, or cylindrical tank treatment system (Figure 8-2). Unlike the red fox systems that are used on most offshore drilling operations, the honey pot does not have a series of weirs and compartments. Its general concept is that of retention time and chlorine treatment. The key to maintaining a successful honey pot sanitary treatment system is to maintain the proper amount of chlorine in the system. Unlike the more sophisticated systems, this can be an extremely difficult task in honey pot systems.

Although the amount of waste that a land based system must treat is a lot less than that of offshore systems, events such as clothes washing and dish washing can throw a system's chemistry completely off. The Npdes requires that certain levels of TSS, BOD and pH are maintained in the effluent. The introduction of one 10 gal. load of laundry or one 5 gal. load of dishwater into the system can change these parameters substantially. The phosphates in the detergents counteract the action of the chlorine. Once the chlorine is ineffective, the BOD increases, the pH decreases, and the TSS increases.

The reverse can be true if very little sanitary waste is being generated. This occurs on land locations where very few personnel are staying on the location. Over-chlorination can create a caustic effluent.

FIGURE 8-2

Honey Pot

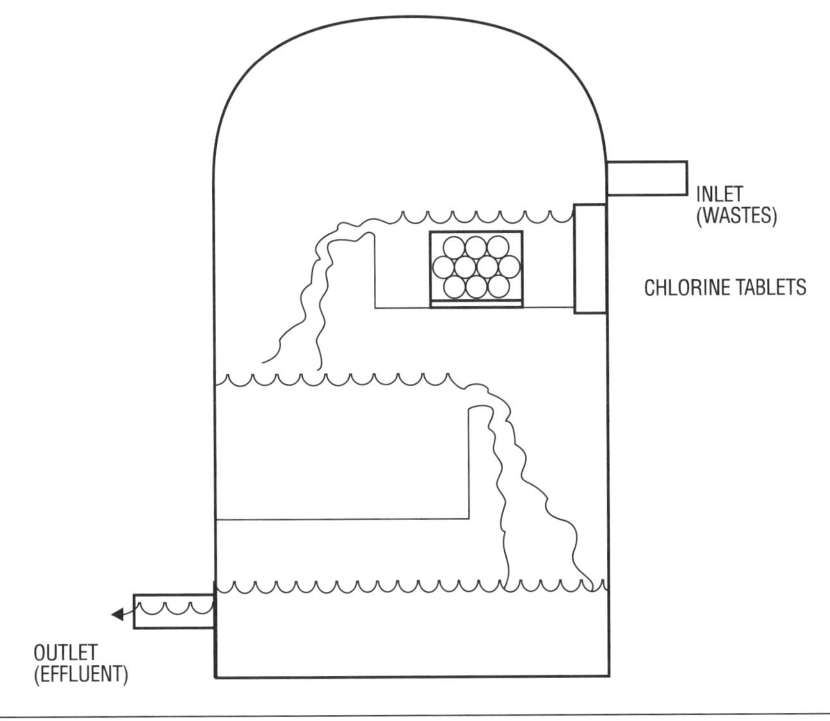

Other than the environmental effects that mistreatment of sanitary waste can have — since on land locations the effluent from the sanitary treatment system is discharged into the rig levee — there is a good chance that the total rig levee fluid may have to be hauled off if it does not meet the Npdes specifications. This can be thousands of gallons of fluid. Therefore it is extremely important that the system on a land based location be closely monitored so as not to contaminate the whole ring levee and risk the costly task of special disposal.

This can be avoided by installing a red fox sanitation unit. Although more expensive, a red fox unit is efficient and requires minimal maintenance. During problem times on a rig, such as kicks, stuck pipe, or other problems, it is sometimes difficult to free someone to maintain the sanitary system. Therefore, utilizing a low-maintenance system may be the best alternative.

Stormwater Runoff

Regulations governing stormwater runoff from oil and gas field locations are proliferating. The regulations are designed to prohibit the unmonitored discharge of any fluids that come into contact with any part of the drilling location. The basis for this regulation is that fluids such as rainwater that come into contact with a drilling operation become contaminated with oil, grease, and other chemicals used during a drilling operation. During a typical South Louisiana rain storm, this can be thousands of gallons of water.

These regulations require that adequate pollution containment devices be installed on and around the drilling rigs to prevent unpermitted discharges of potentially contaminated stormwater onto the surrounding lands or waters.

Lined rig levees can prove to be an adequate containment device. Stormwater that comes in direct contact with the drilling structure itself, including the drilling rig floor, can become extremely contaminated with drilling fluids, oil, grease, and other chemicals. This fluid will then travel from the rig onto the location and then into the ring levee. Although it will have been effectively contained within the confines of the drilling location, it will have potentially contaminated the location from the rig to the rig levee. This means that separate containment devices around the rig floor or rig in addition to the ring levees on a land location would be necessary to minimize contamination.

On an offshore location, since there are no ring levees, the installation of the stormwater collection system is essential. This is especially true on rigs located in fresh water areas, where sensitive wetlands can be directly affected by stormwater discharge.

In highly environmentally sensitive areas such as wetlands or recreational areas, the Npdes rules prohibit the discharge of any fluids into the surrounding environment. This means that all of the gathered fluid must be either hauled off to an approved disposal site or annular injected into the well onsite.

In less environmentally sensitive areas, the discharge of stormwater runoff is permitted if the contractor has obtained an Npdes water discharge permit. The Npdes water discharge permit will usually stipulate parameters such as oil content of chloride concentration before the stormwater can be discharged into the surrounding area. The stipulations that are common on most Npdes discharge permits are:

- No visible sheen.
- No residual oil deposits.
- No residual stains.
- The maximum chloride concentration cannot exceed two times the ambient concentration of the receiving body of water.

While the current cost to dispose of non-hazardous stormwater runoff is not extremely costly, there are regulations minimizing the number of disposal sites and the type of wastes that can be disposed of at these sites. Accordingly, drilling operations plans should include provisions to minimize the properly handle stormwater runoff.

HAZARDOUS WASTES

The classification of a waste that is generated by any oil field operation as hazardous can be extremely costly to the oil and gas industry. Hazardous waste handling and disposal requirements are tenfold more strict than for non-hazardous waste. The special precautions that EPA requires for handling hazardous waste and the limited number of alternatives for disposal of these types of wastes can create an operating environment that is extremely expensive.

Fortunately in the oil industry most of the wastes that are generated are not classified as hazardous wastes. The oil industry enjoys the exemption from the strict scrutiny of the RCRA Subtitle C. Subtitle C specifies the criteria by which a waste is classified and the methods of handling, manifesting, and disposal that are required for hazardous wastes.

Furthermore, the purpose of RCRA is to provide technical and financial assistance for the development of management plans and facilities for the recovery of energy and other resources from discarded materials and for the safe disposal of discarded materials and to regulate the management of hazardous waste.

Drilling and production wastes that currently are exempt from hazardous waste classification under RCRA are:

- Drilling fluids.
- Produced waters.

- Other intrinsic wastes associated with primary production.

But what is not exempt are those motor oils, solvents, and other chemicals that are used on a drilling location. Although the amount of these wastes are comparatively small, the impact of mismanagement is large.

The key for the oil industry to effectively conduct operations under these constraints that RCRA places on drilling operations is to have a thorough understanding of how to minimize the generation of hazardous waste on drilling locations.

When dealing with non-hazardous oil field wastes, often when considering the concentration of contamination, the philosophy of "the solution to pollution is dilution" could be applied. By mixing a non-contaminated waste with a contaminated waste, the concentration can be dropped to an acceptable, non-affecting level.

When dealing with an acutely hazardous waste, however, the dilution approach cannot be used. This is because a hazardous waste is so classified because of the irreversible detrimental effects the hazardous wastes have on life and the environment. When dealing with hazardous wastes, simply mixing a non-hazardous waste with a hazardous waste does not necessarily diminish the toxic effects. Even a small percentage of hazardous waste can have an extremely detrimental effect on life or the environment.

Therefore, under RCRA, the mixing of any amount of hazardous waste with a non-hazardous waste is strictly prohibited. If mixing a hazardous waste with a non-hazardous waste is done inadvertently or wilfully, the total composite waste will then be classified as a hazardous waste and must be handled and disposed of in a manner that is consistent with the rules and regulations outlined in RCRA Subtitle C. This is regardless of the amount of hazardous waste. Even if one quart of hazardous waste is mixed with 20,000 gal. of non-hazardous waste, the total amount then will be considered hazardous waste.

Unfortunately for the oil industry, many environmental activist groups are lobbying for passage of Eckart House bill H. R. 4905 that is designed to eliminate most of the exemptions that are currently enjoyed by the oil industry. A few of the changes that are proposed in the Eckart bill are:

- The elimination of the exemption of associated waters from provisions of Subtitle C.

- Require operators and contractors to remediate abandoned and inactive sites once grandfathered from RCRA.
- Increase fees on drilling and production activities.
- Require the posting of a bond up front to cover any future environmental remediations.
- Increase the requirements on pit lining for drilling and production sites.

These are just a few of the major changes that have the potential of substantially affecting the manner in which oil field waste handling is managed in the future.

The passage of this bill could have a profound economic impact on the U.S. oil and gas industry. Utilizing drilling and production figures from mid-1992, it has been estimated that passage of the Eckart bill could cost the U.S. oil and gas industry more than an additional $1.5 billion/year. This means that thousands of new wells will not be drilled because of the constraints imposed by this bill. With the potential impact of these changes, it is extremely important that drilling contractors have a thorough understanding of what types of wastes are being generated on drilling sites and that mixing of wastes never occurs.

By using new environmentally safe solvents, which for the most part are citrus-based products, the operator drilling or contractor can reduce exposure to the impact of these regulations. The combination of using these products with an understanding of the laws that govern hazardous wastes can save industry millions of dollars.

Naturally Occurring Radioactive Materials (NORM)

When referring to radiation, we often think of nuclear bombs or Three Mile Island. Such considerations lead the general public to become extremely concerned if not sometimes hysterical about the effects of any level of radiation. But radiation need not occur in nuclear accidents or be man-made, it occurs in just about everything we are associated with. Everyday items such as sunlight, the water we drink, and the food we eat all have some degree of radiation associated with them.

The biggest concern is not that radiation occurs, but how we are able to regulate and contain it in quantities that minimize the potential effects to human life. NORM and its relation to the oil and gas industry involves alpha and beta particles and gamma rays that occur naturally in the ground. NORM is transmitted to the surface through drilling and production. Radiation from Radium 226 and Radium 228 are the base forms targeted for regulation by government laws.

Compared with the levels of radiation that are associated with large scale nuclear power plants, the level of radiation associated with NORM is extremely small. If it is extremely small, why is it regulated? The main concern is that inhalation and ingestion of NORM can be extremely detrimental to humans over time. The laws that govern NORM are designed to minimize the potential of inhalation or ingestion by of NORM by human beings.

NORM has existed in the oil and gas industry since the first well was drilled. Most NORM exists as a byproduct of oil and gas production. It exists in production vessels as NORM contaminated formation, as scale in production piping and vessels, and in the fluids themselves. Although not to the same degree as with producing wells, NORM contamination in drill cuttings and fluids is generated during the drilling of wells. Unlike a production operation, roughnecks and roustabouts working the rig floor and the mud tanks have a higher susceptibility to NORM ingestion and inhalation. Therefore it is extremely important that personnel working on drilling rigs be trained in the basics of NORM contamination and how it applies to their operations.

Only a handful of states have regulations that govern NORM. Because NORM comes mostly from oil and gas drilling and production operations, the only states with NORM laws are those with prolific oil and gas operations. These laws govern the surveying, handling, disposal, storage, and treatment of NORM and NORM contaminated materials.

Extremely stringent laws that govern NORM — and in some cases the overreaction of the general public and government agencies about the effects of NORM — usually result from a lack of understanding of the basic principles about it. Unless the oil industry takes the initiative to educate the general public and the government on the basics of NORM, the legal effects from unrealistically structured regulations can be extremely costly to the oil and gas industry. At the same time ignoring the potential problems that can be caused by mishandling of NORM can be extremely harmful to personnel working with and around NORM.

Waste Handling and Minimization

After reviewing the types of wastes that are generated during the drilling of a well and the laws that regulate the handling, treating, and disposal of this waste, it becomes difficult to understand how oil and gas operators and drilling contractors and environmentalists can coexist. On one end of the spectrum companies in the petroleum industry are striving to meet the energy needs of the United States and at the other end, environmentalists are trying to restrict and often prohibit companies from drilling for oil and gas.

Many environmentalists believe that the costs, in terms of environmental damage, of drilling for oil and gas far outweigh the benefits of that effort. But companies involved in drilling for oil and gas believe that the laws inspired by the overreaction of environmentalists are overly stringent and impose severe economic hardships on the U.S. oil and gas industry. This high price tag for complying with stringent environmental regulations has forced many domestic oil and gas companies to dedicate a large portion of their money and efforts towards exploring for oil and gas outside the U.S., where environmental regulations governing the industry are less stringent.

In terms of impact to the environment, much of the environmental effects tied to the oil and gas industry operations do not occur during the drilling of a well. Other than a few acres directly affected as the drilling site, which is restored to its original state after drilling is completed, no other real impact to the environment is felt. Even during the most catastrophic events during drilling, only a minimal amount of damage to the environment can be expected. But this is not to say that inappropriate practices in the past by some companies did not cause sufficient environmental impact to warrant the ire of a few environmental advocates. In fact, unlike today's heavily regulated oil and gas operations, some "get-rich-quick" oil men in the past ran amuck in attempting to drill all they could in a short span of time without much regard to the environment.

Despite the fact that tremendous strides, in terms of technological advances and environmental sensitivity, have been made in the oil and gas industry by major and independent companies alike, government regulations continue to become more prohibitive and costly to the industry. With environmental concerns continuing to grow, this basically leaves only one alternative for the oil and gas industry: to become more environmentally minded in the process of drilling wells.

There are several ways that drilling wastes, such as drill cuttings, drilling fluids, chemicals, et al. , can be handled up front in such a way as to help minimize the amount of waste that is being generated and at the same time allow the drilling contractors to drill wells in an environmentally safe, efficient, and cost- effective manner.

Through the evolution of the oil and gas industry, several techniques and methods for handling drilling wastes have been evolved. These methods range from jury-rigged contraptions to highly engineered electronic and mechanical devices. Many of these approaches, including reserve pit design, land farming trenching and burying, and discharging and reuse require close coordination with the land owners as well as with the regulatory agency to complete successfully. The more advanced techniques, such as subsurface injection, haul-off, closed-loop systems, and slurrification, require technical expertise and training to be successful.

RESERVE PITS

The most widely used method of handling wastes on drilling locations involves the use of reserve pits. On land, reserve pits are earthen pits that are constructed to allow enough area where drilling fluids and cuttings can be stored and segregated for proper disposal after the completion of the well. Offshore, reserve pits are steel tanks designed with enough capacity to store drill cuttings and drilling fluids. Of course, on most offshore locations, drilling contractors are allowed to discharge drill cuttings and drill fluid into the surrounding water. On inland water locations, where environmental laws are becoming more prohibitive regarding the discharge of anything from a drilling rig, a combination of steel reserve pits and discharge usually is utilized.

On land locations where regulations and/or land owners allow the use of earthen reserve pits, it is extremely important that they be designed appropriately so as to not impact more soil than necessary and to ensure that they are accomplishing their required task.

Earthen reserve pits come in a variety of shapes and sizes and have proven to be quite cost effective in handling drilling wastes. It is very important, however, in designing earthen reserve pits that the type and volume of drilling wastes being generated are understood prior to pit construction. The type of fluids, the duration of the drilling job, and the volume of fluids can dictate extreme changes in the way the reserve pit is designed.

The basic design for an earthen reserve pit in the oil and gas industry is a square or rectangular pit. Of course, the size of the pit depends on the type of waste fluid or cuttings that are going to be stored as well as the duration of the storage. Keeping in mind that solids separation is an important part of the pit design, the pit should be big enough to allow proper retention time for adequate solids separation. Proper separation of the fluids allows the solids to be discarded and the fluids (if they meet all regulatory specifications) to either be reused or discharged into the surrounding area.

Although rarely utilized in modern drilling operations, earthen reserve pits used to be utilized as a major component of the drilling fluid circulating system. These pits were engineered to allow enough area and with the right geometry to allow solids separation, thus ensuring that the drilling fluid was clean. This clean drilling fluid then would be reintroduced into the drilling fluid system and used to drill more bore hole. In contrast, today, a series of open top steel tanks are used for this purpose.

Based on the multiple tasks that were required of earthen reserve pits, these pits were quite large. In most cases, the size of the earthen reserve pits was more than 50% of the total drilling location. With the inefficiencies in drilling fluid properties and associated bore hole problems, and with new regulations limiting the size of location that could be used for the reserve pits, using an earthen reserve pit system as part of the circulating system has been all but eliminated.

Even when an earthen reserve pit was not utilized as part of the drilling fluid circulating system, it still played a major role in the storage of cuttings, reserve drilling fluid, and reserve drilling fluid make-up water. The problem that existed with early drilling operations, however, is that only one pit was used for all of these tasks. One pit of 200 ft by 300 ft was excavated, and all cuttings, wastes, and reserve fluids were kept in the same pit.

The problem with this design is that it did not allow for the isolation of contaminated fluid that might enter the pit. In other words, if a salt water or hydrocarbon kick is taken, and salt water or hydrocarbons enter the pit, the whole pit becomes contaminated. With today's environmental regulations, this would mean that the total contents of the pit would have to be treated and/or hauled off. The contaminated fluid cannot be isolated from the rest of the fluid. This can prove to be a costly mistake in the design of a drilling fluid system.

More recent designs of earthen reserve pit systems keep in mind the potential for a kick occurring and contaminating the whole system and attempt to customize a pit system that will fit the individual needs of each well or drilling engineer. These new designs also consider that with stricter environmental regulations, reserve pits with a smaller area while accomplishing twice the task will need to be built.

Recent advances in earthen reserve pit systems utilize a multi-pit system. The multi-pit system allows the use of the same single pit system area and incorporates construction of several smaller pits to accomplish specific tasks. The typical multi-pit system includes a shale pit, a reserve pit, a treating pit, and an emergency pit. Depending on the type of well that is being drilled, additional pits can be added or removed. But this must be done at the time of pit construction. Attempting to modify the pit system after the pits have been constructed and the operation has commenced can be quite difficult.

The shale pit is designed to store the returned shale and sand drill cuttings from the well bore. This pit is usually the largest pit in the system because a substantial amount of cuttings and fluids can be generated during the drilling of a well. The pit also is designed with enough area to allow the solids and liquids to separate. Once the solids and liquids separate, the fluid is transferred to the treating pit and the solids removed via a drag-line (bucket) either to a designated land-farming area or to dump trucks for transportation to a disposal facility. In some cases, the cuttings are left in the pit, slurrified and injected into the well bore annulus.

On a typical 15,000 ft well in South Louisiana, the pit should be large enough to hold cuttings and fluid to about 9,000 ft. Typically, unless abnormal geopressures exist in an area, the cuttings can be spread out on an adjacent tract of land for drying and working into the soil. This is commonly referred to as land-farming.

Beyond 9,000 ft, the drilling rate usually begins to slow down, and the fluid weight is increased by the addition of barite to combat abnormal geopressures in the well bore. Chemicals are added to help the suspension, cleaning, and inhibiting characteristics of the drilling fluid. Once these chemicals are added to the drilling fluid, the option of land-farming usually is eliminated.

The shale pit also is often referred to as the trash pit, because all of the drill solids that are separated by the rig's solids control equipment are discarded into the pit. A substantial amount of fluid also is introduced into the shale pit. In fact, there is typically a 3:1 fluid-to-solids ratio in this pit. Fluids from saturated cuttings, inefficient solids control equipment, washdown water, stormwater runoff, and cellar pump rundown all end up in the shale pit. If not monitored carefully, excessive water usage and excessive rain can cause critical capacity problems with the shale pit. So it is important to consider the effects of not only drilling fluid and cuttings during shale pit design, but also the effects of other sources of potential fluid sources.

One component of the shale pit that probably has given contractors and operators most of their problems in construction and operation of the shale pit is that of the bulkhead. A bulkhead is necessary to allow the drilling rig's steel circulating system tanks and solids control equipment to be positioned as close as possible to the pit. This helps minimize the angle of the "mud-slide," or discharge trough from the shale shakers to the shale pit. By reducing this angle, a smaller volume of water is needed to push the solids from the shale shakers to the shale pit, which in turn decreases the volume of fluid that has to be handled or disposed of. On a typical well, this can equate to thousands of gallons of fluid.

An extremely important factor in designing and constructing the bulkhead is the hydrostatic pressure that the bulkhead is subjected to with a shale pit full of cuttings and fluid. The most common mistake in bulkhead construction is that the pilings are not driven deep enough to contain the pressures. This often results in a structural failure in the bulkhead and the fluids and cuttings spilling onto the drilling location.

Several factors are important in construction of a bulkhead. First, it should be determined how much "free-board," or space from the top of the fluid to the top of the bulkhead, is needed as a safety factor. Free-board length usually will vary depending on the time of year and whether it is rainy or dry.

Next, it is important to have an understanding of the type of soil that the bulkhead pilings are being driven into. If the soil is a sand, then the pilings will need to be driven deeper to allow enough support to contain the hydrostatic pressures. Pre-site soil borings should probably be taken to help identify the type of soils that exist at the site. Third, the type of material that is being used in construction of the bulkhead should be of adequate strength to withstand the hydrostatic pressures of the fluid. The proper size pilings and planks should be used for the retaining wall. The small incremental cost of utilizing the appropriate materials far outweighs the cost of repairing the bulkhead and cleaning up the cuttings and fluids from a deficient bulkhead.

The construction of a bulkhead in a shale pit is not always necessary, however. There are many other methods that can be utilized to keep the mud slide clean, thus eliminating the need to place the shale shakers close to the shale pit. One creative technique to keep the mud slide clean and reduce the amount of fluid added to the system is to install a small pump in the ring levee to draw fluid from the shale pit with a wash down hose connected to the mud slide. Water from the shale pit is pumped onto the mud slide to help keep the cuttings moving. This recycling of fluid reduces the amount of fluid that is added to the system and subsequently reduces the amount of fluid that would have to be either treated or disposed.

In most cases, the decision to install a bulkhead depends on the economic aspects of the pit design rather than the environmental aspects. Therefore, an analysis of the most effective method of shale pit construction should be made. Several factors, such as cost of materials to build the bulkhead versus the cost of pump and hose rentals, should be taken into consideration. Keeping this in mind, proper design of the shale pit can substantially reduce the amount of fluid that must be disposed of at the end of a drilling job as well as minimizing the potential environmental impacts.

The reserve pit is usually the largest earthen pit in the multi-pit system. Its main use in the system is for the storage of either drilling fluid or make-up water. Unlike the shale pit, the reserve pit usually does not have a bulkhead.

The drilling fluid that is stored in the reserve pit is usually at the same density as the drilling fluid being used in the current bore hole section of the well or at least within 0. 2 lb/gal. so that only a minimal amount of weighting material will need to be added to bring the fluid to current drilling fluid density. This is usually desired so that in the event that a salt water or hydrocarbon kick is taken or if severe lost circulation while drilling is encountered, the fluid will be readily available to help combat these problems.

Water stored in the reserve pit is usually utilized as make-up water for the drilling fluid. The reserve pit water also is often used as hole fill-up water in the event that severe lost circulation in the well bore is encountered. As much as 10,000 bbl of water can be stored in a reserve pit. The water source for the reserve pit is usually a water well that is located at the site. If a water well is not located at the site, then the water is usually trucked in.

If the fluids during the drilling of the well are managed properly, the amount of reserve pit water remaining at the completion of drilling will be minimal. The smaller the volume of fluid that remains in the reserve pit, the less the cost of having to dispose of the excess reserve pit fluid, either through injection or haul-off, at the end of the job.

A treating pit is increasingly common on most drilling locations. With environmental regulations becoming stricter concerning the discharging of any fluid from a drilling location and the cost of hauling the fluid to a disposal site escalating, the concept of maintaining a treating pit on location to chemically treat contaminated fluids onsite for discharge or reuse purposes is rapidly gaining acceptance.

The treating pit can be as large as a reserve pit, depending upon how much fluid is anticipated or desired to be treated. Contaminated fluids from other pits in the system or from the well are pumped into the treating pit. The contaminated fluid is then tested for the type of contamination that exists. A chemical treating program is then formulated and applied to the fluid. The chemical treating program is dependent upon what type of contaminant is present in the contaminated fluid and to what parameters are desired. If it is desired to reuse the fluid as make-up fluid or drilling fluid, the treatment can be fairly straightforward. If it is desired to bring the fluid to discharge parameters, then the treatment becomes more difficult and costly.

Any fluid that is going to be discharged must meet Npdes standards. Depending upon which state the well is being drilled in, the standards may fall under federal jurisdiction or state jurisdiction.

The cost of treating any contaminated fluid is not inexpensive. But compared with the cost of having to haul off all of the contaminated fluid to an approved commercial disposal facility, the cost of treating is reasonable. If discharging from the pit to a land farming area is desired, then the pit should be situated such that it can be accessed easily by pumps and tanker trucks.

One pit that may never be used throughout the drilling of the well, the smallest but sometimes most cost-effective pit in the multi-pit system, is the emergency pit. The emergency pit is usually placed near the shale pit and close to the drilling rig. This allows easy access to the pit in the event that contaminated fluids need to be diverted to the pit.

The utilization of an emergency pit on a location minimizes the amount of severely contaminated fluid that needs to be treated or hauled off. Unlike a single pit system where if a kick or cement is returned the whole pit is contaminated, in using an emergency pit, only a very small volume of emergency diversion of well bore fluids will need to be treated or disposed.

On almost every well, some type of contaminated fluid is going to be created during the course of drilling and completing. A typical contaminated fluid that is generated is cement. During the surface casing job, cement returns are always desired to ensure that cement has covered the entire length of the casing. The amount of cement returned can vary depending on how much actual washout has occurred during drilling and if the cement has "roped" during displacement. Once cement returns have been noticed, the return fluids then should be diverted into the emergency pit and the cement immediately treated with sugar to retard its setting time and allow it to be safely hauled off.

In other instances where a hydrocarbon or saltwater kick is taken and the contaminated fluid is brought to surface, the emergency pit can be used to store the returned hydrocarbon fluids or saltwater. As mentioned before, this can save tens of thousands of dollars in additional treating and haul-off costs if the whole system were contaminated by these fluids.

Therefore, it is critical on any land location where earthen pits are going to be used, that consideration be given to utilizing a multi-pit system. The multi-pit system allows modular treatment of all fluids generated, used, or stored on a drilling job. In the long run, any cost to design and manage a multi-pit earthen pit system can be easily offset by one incident where the total pit is contaminated by a small amount of contaminated fluid.

The storage and treatment of fluids offshore is less cumbersome than on land. Just as with onshore locations, it is important to maintain a certain amount of reserve fluids on location in the event of an emergency. Unfortunately, the amount of reserve fluid that is stored is limited due to space and weight limitations on the platform, jack up or semisubmersible drilling rig.

Fortunately, unless a drilling contractor is utilizing a toxic or oil based drilling fluid, federal regulations allow the discharge of drill cuttings and drilling fluids overboard into the surrounding body of water. The discharge fluid does have to meet certain criteria, however.

In federal waters, where the Minerals Management Service (MMS) maintains jurisdiction, the discharge fluid, whether it be drilling mud, water or cuttings, must be tested and pass the Mysid shrimp mortality test, or as it is formally known, the LC-50 test. This test is a measure of the toxicity of the fluid as measured by the 50% mortality of the specially raised Mysid shrimp. A contractor is required to have the fluid tested once per week. If the fluid fails the LC-50 test, the contractor must immediately rectify the problem and can be subject to fines as high as $10,000/day, retroactive to the last date of a successful LC-50 test.

In most cases, this is not a problem because the test is designed around the typical constituents that are used in drilling fluids. If the drilling contractor is utilizing a known toxic drilling fluid or an oil based drilling fluid, then no discharge of fluids will be allowed, and the drilling contractor must employ a closed loop drilling fluids system.

When drilling a well on inland waters, or in most cases wetlands locations, the regulations governing the handling, storage and discharging of fluids involve a combination of upland and offshore techniques.

In most inland water locations where water depths are less than 12 ft deep, a drilling barge is utilized. Unlike land locations, no earthen pits are used for fluid storage, and unlike an offshore drilling location, discharging of fluids is not allowed. This leaves but one alternative for drill cuttings and fluids storage and handling:closed loop systems. What is fortunate on an inland water location is that cuttings barges and fluids barges can be placed at the location to store cuttings and fluids. This method of handling cuttings and fluids, in combination with the closed loop system, can be effective.

On inland water locations where some discharge of fluids is allowed, the fluids must be tested for certain wastewater discharge parameters. In no case with today's wetlands regulations, is it allowed to discharge drilling fluids or drill cuttings into a marsh. This regulation is governed by the sensitive ecology found in coastal and wetlands areas.

LAND FARMING

As was touched on upon briefly, simply discharging returned drilling fluid and drill cuttings under environmental regulations can be difficult. Nonetheless, if the well is located in an area that allows land farming of these wastes, it can save the contractor or operator thousands of dollars in handling and disposal fees per well.

Land farming basically entails the removal of the drill cuttings and drill fluid from the shale, reserve, and treating pits and spreading them on a tract of land that is 1 ac or more in area. The dried cuttings then can be tilled into the soil and farming of the property can continue as if nothing had changed. In most cases, the land farming of these cuttings and fluids will increase the yield of the crops grown on the land farmed area because most fluid that is allowed to be land farmed is extremely rich in nutrients and thus acts as a fertilizer.

Applying the land farming technique also is relatively inexpensive. The only equipment that is required to implement a land farming operation is a drag-line and a bulldozer. Depending upon where the land farming site is situated in relation to the pit system, a dump truck may be required to transport the cuttings and fluid to the land farming site.

Before initiating a land farming operation, however, three major hurdles must be overcome. First, permission must be granted by the land owner to allow the contractor to land farm any fluid on his property. This becomes difficult if the particular land owner in question has had a bad experience in past with land farming by another drilling contractor. This is usually handled by a petroleum landman at the time the lease for the site is being negotiated. It would probably enhance the process and improve prospects for obtaining this permission if a person knowledgeable with the process, such as a drilling engineer or drilling foreman, meets with the land owner and landman at the time of negotiation.

Second, at the same time or before negotiations are taking place with the land owner to allow land farming, the contractor or person in charge of obtaining the permits for drilling the well should seek approval from the appropriate agency that grants land farming of oil field drilling fluids and drill cuttings. Depending upon the state that the well is located in, this could be the Department of Natural Resources, the Department of Environmental Quality or the U.S. Army Corps of Engineers. It would not do any good to obtain permission from one party, the land owner or the agency, if the other disallows the operation.

Finally, before commencing the operation, an estimate of the amount of drilling fluid and drill cuttings that is going to be generated during the drilling operation should be made. This is extremely important in the shallow portion of the well (down to 9,000 ft) because land farming of drilling fluids and drill cuttings beyond a certain depth in the well is not allowed, and the amount of drill cuttings and drilling fluids generated in the shallow portions of the well account for two-thirds of the waste in the entire well. If the amount of waste is not estimated, then the land farming area and pits could be grossly under-designed or over- designed, thus causing waste.

It is important to note that just as with any drilling-related operation, pre-operational samples of the land farming site should be taken. This will help explain any problems that may occur on the land as result of the land farming operation and in the event the farmer yields higher returns as a result of the land farming operation.

At the end of the land farming operation, which is governed by legal limitations or fluid properties, post-land farming samples should be taken and laboratory tested in the same manner and for the same parameters as the pre-land farming operation. This will ensure that the proper mixing of the soil has been done and the parameters are within the legal regulatory limits. These analyses also provide the drilling contractor and operator with evidence that the soil was not contaminated in the event a problem with the land develops later and the land owner is seeking compensation for damages because of reduced crop yield or dead vegetation. This is not to say that this is an absolute guarantee against any future repercussions or claims, but it provides the contractor and operator with a defense against any potential unfounded claims.

Trenching and Burying

Trenching and burying of wastes can be a viable alternative for the disposal of some wastes is generated by drilling operations as long as the waste is not considered hazardous and ground water is not affected. Much like land farming, trenching and burying entails leaving wastes at the site in a non-damaging form. For an even better analogy, trenching and burying resembles techniques that are utilized in construction and demolition debris landfills.

As with construction and demolition landfills, a trench is excavated to a depth just above groundwater, and non-hazardous wastes are placed in the excavation. A cover of natural soil then is placed on top of the waste. In drilling operations, the waste can be shales, clays, and sands and even wood boards that were utilized for the drilling site. What is not allowed are items that are considered hazardous, wastes that emit odors, or wastes that have a high propensity to leach contaminants into the groundwater. If the area geology allows, the utilization of trenching and burying can save a contractor thousands of dollars in haul-off and disposal costs.

There are a few items that should be investigated before trenching and burying can or should be pursued. First, before the lease is even negotiated, the drilling engineer, drilling foreman or geologist in charge of handling the environmental aspects of the well should investigate the shallow geology of the site. This should include the type of soil that exists at the site and the depth of groundwater. If the soil type is sandy or silty and leaching mobility is high, and if the groundwater depth is shallow where the trench will penetrate it or is extremely close to the bottom of the proposed trench, then the drilling contractor or operator should reconsider the use of trenching and burying as a method of waste disposal.

Next, as with land farming, prior to considering trenching and burying, consent from the land owner and the appropriate state agency should be obtained. Although no wastes detrimental to the environment will be buried, a permanent physical change to the land is being proposed. Regardless of what positive economic effects trenching and burying may have, if prior permission is not obtained, both the land owner and agency can press claims for wrongdoing. This can best be handled during the lease negotiations, much as with land farming. The employee responsible for this portion of the operation should be a vital part of the lease negotiations, including explaining the positive and negative aspects of the proposed operation.

Another important aspect of the process is the identification and amount of waste that is being generated. This is important because trenching and burying is usually utilized as a supplement to disposing certain types of wastes, not as the sole method of disposal. Items that are usually allowed in a trenching and burying operation are non-contaminated location boards and non-hazardous drill cuttings such as shales, silts, and sands. If a large volume of these types of wastes is expected to be generated at the site, then plans should include using a combination of trenching and burying and another technique to dispose of the wastes.

As with any operation that involves the alteration of natural or current environmental characteristics by the introduction of an oil field waste, pre-site and post-site sampling and analysis should be conducted. This will help safeguard the drilling contractor or operator against any future claims of the operation damaging groundwater, soil or crops. Depending upon the size of the trench, at least two samples should be taken and analyzed for any contaminates.

One common approach for utilizing trenching and burying at a location call for the buried mixture to be at least 5 ft below ground level and covered with 5 ft of native soil and at the bottom of the trench be at least 5 ft above the seasonal high groundwater table. Such criteria are intended to prevent the cultivated portion of the trench (i. e. , from surface to 3 ft) from being contaminated and the groundwater from being contaminated.

In any case, the use of trenching and burying can serve as a suitable alternative for disposing of several types of wastes generated on a drilling location. If coordinated and applied properly, trenching and burying can be beneficial to the drilling contractor, land owner, and environmental agency.

Discharge

The discharge of drilling wastes as a method of handling and disposing of them basically is a combination of several types of methods. On land locations, discharging other than into a pit where the wastes either will be hauled off, land farmed, or injected is extremely rare. Generally, discharge of drill cuttings and associated drilling wastes has not been conducted in oil and gas drilling operations since drilling contractors and operators have become more environmentally conscientious.

On inland water locations, other than a few special cases, such as treated stormwater runoff, direct discharge is strictly prohibited.

Discharge is basically limited to offshore rigs that operate in federal waters. As was discussed before, drilling fluids discharged from a drilling rig in federal waters must meet criteria dictated by the MMS. Weekly laboratory analyses on the drilling fluids are required to ensure that the criteria are being met.

There are two basic criteria that the MMS requires of a fluid that is being discharged from a drilling rig offshore: a sheen test and a Mysid shrimp mortality, or LC-50, test. When a contractor discharges drilling fluid or drill cuttings overboard, he must inspect the discharge for the sign of a sheen that is caused by hydrocarbons. If a sheen is noticed, the law requires that the discharging of that waste be discontinued until the problem is rectified and should be reported to the MMS. If it cannot be controlled, then the contractor is required to go directly into barges with the wastes or install a closed-loop system on the rig to catch the wastes.

Every week, the drilling contractor is required to obtain a sample of the drilling fluid and have it laboratory analyzed for LC-50. The LC-50 tests the drilling fluid for its toxicity. This is done by determining the mortality rate of the Mysid shrimp in a diluted solution of the drilling fluid for 94 hr. If the mortality rate of the Mysid shrimp is greater than 50% in the test period, then the drilling fluid is considered toxic and discharge of the fluid must be halted. If the sample exceeds the LC-50 limit, the contractor must correct the problem and retest the drilling fluid. If the problem persists without correction, the contractor is subject to a fine of $10,000/day retroactive to the last day of a successful test. The contractor and the laboratory are required to submit the test results to the MMS for review. If it is determined that the drilling fluid cannot meet the discharge requirements, it either must be changed out or a closed-loop system installed.

Under no circumstance is it allowed to discharge oil-based drilling fluids or toxic drilling fluids. In most cases, drilling contractors utilizing oil-based drilling fluids will employ a closed-loop drilling system. This should ensure that no fluid is being discharged into the water.

There is another situation that sometimes forces a drilling contractor to cease the discharge of drill cuttings and drilling fluids into the water. This is when stuck pipe occurs and an oil-based spotting fluid such as Black-Magic is used. Many offshore wells are drilled directionally from a platform. With the advent of sophisticated drilling methods such as top drive drilling motors, measurement while drilling instruments, and aluminum drill pipe, horizontal displacements of more than 10,000 ft and drilling angles of greater than 75° are common. These drilling techniques allow the exploration and development of geologic plays that would otherwise be left untapped. But along with this advanced technology, the probability of stuck pipe is increased. When stuck pipe occurs, it is extremely important that every available means to free the pipe is employed as soon after the incident as possible. There is an exponential correlation between time and permanently stuck pipe.

Of all the products and techniques used in the oil field to free stuck pipe, the application of a spotting fluid has proven to be the most successful. The most successful of the spotting fluids utilized in the industry is an oil-based one called Black Magic. Black Magic is designed to dissolve the wall cake that builds up on the side of the bore hole that the drill pipe is stuck to. If applied within hours of when the pipe is stuck, Black Magic has an extremely high success rate in freeing the stuck pipe. Conversely, if Black Magic is not used at all or if a large amount of time since the stuck pipe has elapsed, a much lower success rate in freeing the stuck pipe can be expected.

Once the Black Magic has been applied and the pipe is freed or the spotting fluid circulated out, the returns must be captured on the rig, in a barge, or in a closed-loop system. In most cases, the amount of spotting fluid used is only a fraction of the total amount of drilling fluid in the drilling system. In this case, the spotting fluid can be incorporated into the system with little to no adverse effects to the system. Once testing of the fluid indicates that there is no sheen from the incorporated drilling fluid and the LC-50 test is positive, then once again the drilling fluid can be discharged. Larger amounts of spotting fluid must be removed from the mud system, however.

The same rules and regulations governing the discharge of oil-based drilling fluids also govern other exotic drilling fluids such as salt-based or specialty chemical-based drilling fluids. By being aware of the area in which the well is being drilled, the type of fluid that is being used, and the regulations that govern the discharge of the fluids, a contractor can be successful in minimizing costs of waste handling on any drilling location.

RE-USE

The concept of re-use of materials is not limited to recycling aluminum cans and newspapers. The concept of re-use or recycling has also made its way into the oil and gas industry. The re-use of location materials, drilling fluids and drill cuttings all have been introduced in the drilling industry.

On many drilling locations, thousands of board feet of boards are used to construct a drilling location. With the extremely heavy loads that the boards are subjected to, most of the time the boards are disposed of after each job. This can be an extremely large amount of boards per well. The concept of modularization has been introduced into the industry. Basically, just as with parquet flooring, mats of varying types of plywood can be constructed that are more resistant to wear and easier to install. Because the mats are more resistant to wear, they can be used over and over on multiple locations. Furthermore, if one mat is damaged due to an unusually heavy load, the individual mat can be replaced without dismantling the whole location. This cuts down on the amount of boards used as well as the cost of constructing drilling locations.

For most in-field drilling operations, the geology and paleontology remain consistent from well to well. Casing points and weight-up intervals can be selected based on drilling depth versus the analysis of geology or paleontology. When this is the case, the drilling contractor has the opportunity to substantially reduce the cost of drilling fluids for each well. Merely processing the drilling fluid in the solids control system can condition it to be used on multiple wells. In non-geopressured drilling areas, the drilling fluid can cost as much as several dollars per foot drilled. This includes build-up and maintenance of the drilling fluid. If an in-field drilling operation is planned in a non-geopressured area, this cost can be reduced to well below $1/ft drilled.

But re-use is not viable in all areas, however. In areas where it is critical to select a casing point or a geopressure point it is not recommended to utilize used drilling fluid. The paleontological markers that are used to select these points become very critical. If a used drilling fluid is utilized, the paleontological markers from a different zone may be mistaken for a desired marker and a casing point or geopressure point missed. This could serve to be extremely costly in terms of money and bore hole problems.

In any case, each well and field should be reviewed for the application of such cost and material savings techniques. For instance, if a well is drilled in a field that has more than 300 wells, but the field is situated around a salt dome, this probably is not an area where used drilling fluid should be used. This is because of all of the faulting that occurs around a salt dome, thus making each casing point or geopressure point different in every well. But in a West Texas, flat geology, infield drilling program, this method could serve to save hundreds of thousands of dollars over the life of the drilling program in the field.

Recent efforts have been made to re-use drill cuttings from a drilling well as road bed material. One well generates thousands of barrels of shale and clay. If processed properly, these cuttings can be utilized as base material on road beds.

The process includes drying out the cuttings in large furnaces at elevated temperatures, leaving the dried shale and clay in a form that can be crushed and used as road-bed material. This is much the same process that is used in the making of Portland cement. The raw material, which is mainly shale and clay, is fed into a kiln. The temperature within the kiln is at or about 2,000° F. The moist shales and clays are dried until no traces of water exist. The dried material is then crushed by grinders and crushers, and the end product is used as roadbed material in the construction of roads. The size of the final grind is usually less than 2 in. diameter spheres.

This same process has been used not only for non-hazardous oil field wastes but also for hazardous wastes. The same process is utilized in attempting to burnoff the toxic constituents of the hazardous wastes.

One major concern of utilizing the kiln techniques in processing hazardous wastes involves the toxic emissions that may be emitted as a result of the drying process. Much like an incinerator, the kiln attempts to dry the product at a controlled temperature. Because the temperature is maintained at about 2,000° F. , emissions into the atmosphere still can have a certain level of toxicity. This airborne contamination can be inhaled or ingested and cause harm to life. This is one reason that utilizing the kiln process to dry out hazardous wastes should given careful consideration. Many companies that propose this type of hazardous waste handling technique are finding numerous legal roadblocks in getting these types of processes approved and accepted.

One of the recent concerns in the use of drill cuttings for roadbed base is that of the potential presence of NORM in the cuttings. Because the clays and shales are naturally occurring and NORM is generated naturally, much of the drill cuttings that are generated by the drilling process can contain NORM. Although the kiln or incinerator process may be partly successful in minimizing the amount of hazardous constituents that remain in a waste after processing, it still is not clear if the incinerating process minimizes the amount of NORM that may be contained in a waste.

Another method that reduces or attempts to reduce the amount of NORM in a soil is that of the multi-cell treatment system. This technique incorporates the idea of dilution to reduce NORM levels. The waste is processed in several stages by mixing the NORM contaminated material with fresh, non-NORM contaminated material. This dilution is continued until an acceptable level of NORM is maintained in the material. The final product can be returned to the owner or used as roadbed material. But much like most recycling processes, this is a fairly new technique and is not recommended until all state agencies and the EPA approve the processes.

SUBSURFACE INJECTION

The use of the disposal option known as subsurface injection allows the contractor to dispose of the drilling wastes without affecting the surface environment or haul it off to an offsite disposal facility.

Subsurface injection involves the pumping of drilling wastes, (drilling fluid, drill cuttings, et al.) either into the annulus of the well or if the well is a dry hole, down the intermediate casing into a zone or zones that are isolated from potable water supply sources and hydrocarbons. Even if a well is considered a dry hole, it is usually preferable to utilize the annulus for subsurface injection of wastes.

The only criteria that are required to conduct subsurface injection is that permission be granted to the contractor by the land owner, that a permit is obtained from the appropriate government regulatory agency, and that the particle size of the waste will allow it to pass through the injection points.

As with any operation conducted on a lease, prior to the contractor conducting any waste disposal operation on the property, permission either through the lease agreement or via another written legal form should be obtained. On offshore locations, this is usually not a problem because no land has the potential to be affected. It is also important to note that while there is no immediate threat to any usable real estate that the land owner possesses, if there is a breakdown in the formation and the waste broaches to surface, there could be some environmental damage.

In most states, an injection permit first must be obtained from the appropriate regulatory agency before any injection can take place. The agency basically ensures that the well has the integrity to withstand the injection pressures that process will impose and that no critical resources, such as drinking water, will be affected by the operation.

The most critical factor that can affect the subsurface injection process is that of waste particle size. Since the casing perforation size is limited and the annulus size is limited, the size of the particles should be carefully controlled. In most cases the drill cuttings are ground in a process called slurrification. The concept of slurrification will be discussed in more detail elsewhere in this book.

The ground cuttings are mixed with water and a suspending agent and sieved to ensure a particular particle size. The waste mixture is then pumped from a holding tank either into the intermediate casing string or into the casing-to-casing annulus (i. e. , between the surface casing and the intermediate casing). Because the area of the perforations is limited and/or the space between the surface casing and intermediate casing is limited, particle size of the waste is extremely important. Particles that are too large could plug off the perforations or bridge off in the casing-to-casing annulus. This could force the contractor or operator to abort the injection operation. Proper grinding and screening of the wastes before they are injected is essential.

Before selecting subsurface injection as an option for disposing of drilling wastes, several factors should be explored. By not properly analyzing the mechanical and economic feasibility of this operation, it could cost more than other methods of disposing of the waste.

The first factor in deciding to utilize subsurface injection is whether or not there is a well with the appropriate mechanical integrity available for the operation. In most cases the well that is currently being drilled or where the waste was generated is available for this operation. In more developed fields, the use of an offsetting well may be assessed for this operation. Once it is determined that a well is or will be available for subsurface injection and all legal issues for using the well have been negotiated, the mechanical integrity of the well should be investigated. Here is a list of items that a potential subsurface injection candidate should be investigated for:

- Amount of waste to be disposed of.
- Well accessibility or proximity to waste.
- Wellhead integrity, including operating valves, casing, and appropriate casing slips.
- Adequate drinking water zone isolation.
- Injection zone characteristics (e.g., fracture gradient, size and length of zone, et al.)
- Review the historical success of subsurface injection in the subject area.
- If a producing well, adequate producing zone protection.
- Review the cost effectiveness of the proposed subsurface injection project with other disposal alternatives.

Before selecting or discarding the subsurface injection option for a particular operation, the amount of waste should be estimated. If the quantity of waste is not large enough, it may not be an economically viable option to utilize subsurface injection. If too much waste must be disposed of, the contractor may run the risk of the injection zone capacity being reached or a mechanical failure in the well occurring. The estimation should include both the amount of waste that has been or will be generated from drilling the well and the increase in the amount of waste that will result from the slurrification process.

One factor that is often overlooked or not thought of initially is that of the proximity of the waste to the injection well. It should be obvious that if a well currently being drilled is going to be utilized for the operation, accessibility to the waste must be easy. The waste pits that are storing the wastes are typically within 50 ft. of the injection well. But if the proposed injection well is located away from the waste site, the cost of loading and trucking the waste to be injected may render the injection project economically infeasible. Realistically, any well within 100 ft of the waste site can be considered for injection.

The condition of the well bore that is going to be used as the injection well is one of the single most important factors of deciding whether or not to utilize a well. This is due to the pressures and the abrasiveness of the material that is going to be injected. The casing valves, the casing, and the casing slips all must maintain a certain amount of integrity before the well should be considered as an injection well. None of these items is less important than the other.

The casing valve(s) are the first and last defense of the wellhead in the event problems occur in the operation. All casing valves should be properly inspected and tested to ensure that they will meet the worst case conditions of the operation. If through this inspection and testing, it is determined that the valves do not meet the minimum criteria, then if possible they should be repaired or replaced. If this cannot be accomplished, then the utilization of that particular well bore for subsurface injection should be carefully reviewed. If the valve cannot be replaced or repaired, the contractor can also opt to "piggy back" the deficient valve with a new valve. This will allow minimal work to have to be conducted to get the wellhead in appropriate working order. The biggest drawback is the cost of a new or reconditioned piggyback valve.

One of the biggest problems that can occur during any oil and gas field operation is that of worn or damaged casing. Since casing is the contractor's protection from the bore hole pressures and fluids, a hole in the casing can doom any well.

Before a well is considered as an injection well, the contractor should pressure test the casing and run a casing inspection log. Common practice calls for testing the integrity of the casing (and the connections) to about 80% of the published collapse or burst pressures. If the waste is going to be injected down the inside of the casing, then the burst pressure integrity of the casing is extremely important. During the drilling operation, drill pipe rotation can cause severe wear in the casing, thus reducing the pressure integrity of the casing. Even on the most controlled injection projects, uncontrollable pressure spikes can cause failures in the casing.

Prior to deciding if the casing in a well will withstand the pressures of the injection project, a multi-finger casing inspection log should be run. A multi-finger casing inspection log will show areas of the casing that have been thinned as a result of casing defects or drill pipe rotation. Even a reduction in casing wall thickness of 0. 01 in. in combination with the other radial, tangential, and tensile forces that the casing is subjected to can cause a casing pressure failure. Typical areas of wear in a well are in known "dog-leg" areas, near the surface and at the casing slips.

Since even the straightest wells are not straight, the casing and drill pipe will be subjected to bends (dog-legs), in the well. This basically means that there will be a low side on the casing, and rubbing between the drill pipe and casing will occur. In directionally drilled wells, this type of wear can be substantial. Since it is next to impossible to control dog-legs, the contractor should always plan on running a casing inspection log before conducting a subsurface injection project.

A lot of wear in a casing string occurs at or near the surface. This is mainly a combination of pipe rotation and tripping and the stresses imposed by the casing slips. With as much as a million pounds of weight being imposed on the casing, some crushing of the casing can take place. This crushing reduces the wall thickness and subsequently reduces the burst and collapse resistance of the casing.

When drilling wastes are being injected into the well, a great deal of abrasion to the wellhead equipment and the casing can occur. If annular injection is going to be used, it is important to understand the mechanics of the injection process. At the point where the fluid and/or waste is being injected, a sand-blasting type action at elevated pressures is occurring. If the pressure is elevated enough, the waste contains enough solids, and enough waste is being injected, the point of impact on the casing through the casing valve can be washed out. Once this washout occurs, the waste fluid will enter the well bore instead of pumping downhole into the proposed injection zone. If this is a non-producing well, this is not a big problem. The contractor can simply stop pumping the waste into the well and attempt to patch the casing. If the well is currently on production, and if there is pressure on the tubing-casing annulus, via a tubing leak or otherwise, this washout could cause an uncontrollable blowout of the well. In this case all of the efforts to cut costs in the disposal of the wastes are eclipsed by a new problem.

There are basically a couple of ways to help avoid this happening on a well. Both methods involve the planning of disposal before the well is drilled. The cost to implement these methods, even if the well is eventually used on a disposal well, is minimal, so very little money will be wasted. The most desirable technique to protect the casing in an annular injection project is to install a carbide skirt on the bottom of the casing slips (Figure 9-1). The skirt can be attached to the bottom of the casing slips with minimal effort and at an extremely low cost. The skirt also does not change any of the operating characteristics of the slips. The skirt is designed to take the impact of the injecting fluid and divert it downhole, thus eliminating any chance of washing out the casing. Consideration as to the type of material that the skirt is made of also should be given. If the drilling waste contains abrasive materials such as hematite in it, it is important that the skirt be made out of a material that can withstand the impact. In this case a carbide coated or carbide skirt should be used. If non-abrasive drill cuttings or wastes are going to be pumped, then a standard steel skirt can be used.

If it is felt that the area of impact will be erratic or larger than a skirt can divert, then a long piece ofcasing in place of the skirt should be used (Figure 9-2). The casing can be attached to the bottom of the casing slips in the same manner as the special skirt can. The length of the skirt can range from 1-2 ft. in length to as much as 30-40 ft. in length. Depending on the abrasive characteristics of the waste fluid, various casing grades can be used. Casing grades of J-55 or N-80 can be used for less abrasive wastes, and casing grades of P-110 or Q-150 can be used for the more abrasive waste fluids. Like the skirt, the long casing sheath is installed at the time the casing slips are set.

A major environmental concern that arises with a subsurface injection project is how best to protect underground drinking water resources from contamination. Drilling regulations usually dictate the amount of coverage that is required when a well bore penetrates and passes through a drinking water bearing zone. General regulations dictate that the water bearing zone be covered by at least 200 ft. of casing below the zone and isolated with cement below, and at least 200 ft. above the zone.

The agencies that enforce this regulation look at two pieces of information to ensure that adequate zone isolation has taken place. The first documentation to be reviewed are cementing records. The cementing records are reviewed to ensure that at the time the casing was set, enough cement was pumped to give proper zone coverage. If the theoretical calculations of annular space plus washout do not indicate that there is enough cement to properly cover and isolate the water bearing zone, a cement squeeze job may be required if the well is to be used as an injection well. Another sure way of determining if the cement that was pumped during casing setting is to ascertain if cement returns were realized during the cementing job. If no cement returns were realized during the cementing of the surface casing, either not enough cement has been pumped, the washout around the casing is larger than anticipated, or the fracture pressure at the casing shoe was not adequate to hold the column of cement and lost cement returns were realized.

If it is unsure which of these occurred, and if records indicate that there is a possibility that the water bearing zone(s) of interest has not been adequately covered, then the regulatory agency may require or the contractor may choose to run a cement bond log in the well. A cement bond log is designed to show the percentage of cement bond between the casing and the formation. The complete cement bond log requires only a few hours to complete, and the contractor and the regulatory agency now will have a positive indication of whether or not the water bearing zone has been adequately isolated. Furthermore, if the cement bond log indicates that the zone has not been adequately isolated, the cement bond log can be used to determine where a squeeze job should be applied.

If a cement squeeze job is required to create adequate zone isolation, the cost of utilizing the well as an injection well should be reviewed along with other disposal alternatives. The squeeze job entails the cost of the cement, the cost of perforating for the squeeze, the cost of drilling out the inside of the casing after the squeeze job and the cost of running another cement bond log. This can run to tens of thousand of dollars.

Even if all of the other parts of the well indicate that the proposed well meets all criteria as an injection well candidate, if the formation that is targeted for injection cannot be easily injected into or is inadequate in size, the well may be doomed as an injection well.

During drilling of the well, an idea as to the fracture gradient or to the capability of the formations in the well to take fluid was probably determined. The key to disposing of thousands of barrels of fluid into the well is the injection zone's ease of taking fluid. If the fracture gradient in the well bore is not known or was not determined during drilling of the well, the contractor should run an injection test on the targeted zone. This basically entails filling the casing with fluid, perforating the interval, and determining the pressure required to break the formation down. If the zone refuses to break down or is resistant to taking the fluid, the use of that particular well for waste disposal should be reconsidered. If the particular well is the only alternative for disposal in the area, and if the formation does not break down or is not easily injected into, a hydrochloric acid job to assist in the formation job may be used. In many cases, the sands are intermixed with clays that impede the flow of fluid. The acid will dissolve the clays and allow the formation to take fluid more easily.

Care should be taken during the breakdown of the formation so that the limitations of the well hardware, e.g., wellhead, christmas tree, or casing, are not exceeded. If the integrity of the well hardware will allow the pressure to be increased to at least 80% of the published yield pressures, then it should be attempted. If it does not, however, the acid job or other methods for formation breakdown should be employed.

The size and the length of the zone also are extremely important in selecting an injection zone. On most wells, the intermediate casing hole section is usually logged for geological tie-in purposes. This electric log, which is made up of a gamma-ray, spontaneous potential, and resistivity log, will detail the length of the proposed injection zone. By knowing the length of the injection zone, the proper amount of perforations can be

FIGURE 9-1

CARBIDE SKIRT ON CASING SLIPS

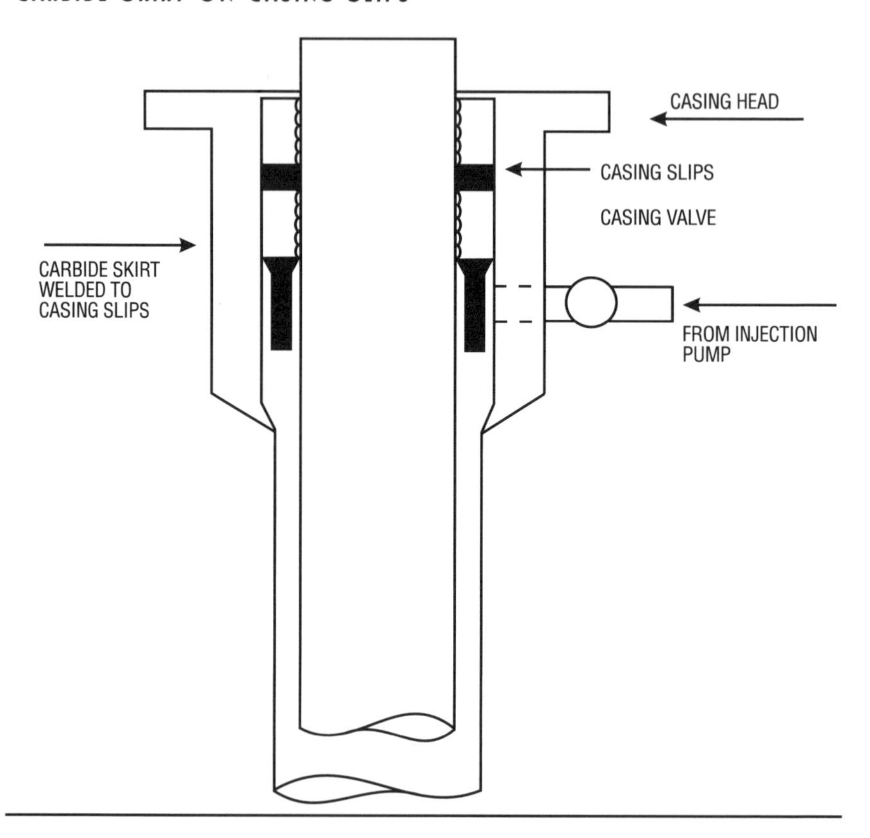

planned to help minimize the amount of surface pressure, caused by friction pressure, that will be encountered during the injection job. When correlated with offset well data, the size of the proposed zone can be estimated. If the amount of waste that is to be injected exceeds the estimated zone capacity, then the contractor may want to search for another zone in the well bore. If an intermediate hole section log is not available on the proposed injection well, then offset or regional geology may need to be reviewed to determine zone characteristics.

Due to the amount of potential problems that can occur as a result of an injection project, before a subsurface injection project is to be explored for a particular well, a historical review of the success or failure of subsurface in the area should be made. The contractor should review any and all available data that may indicate whether or not proceeding

FIGURE 9-2

Long Piece Of Casing Attached To Casing Slips

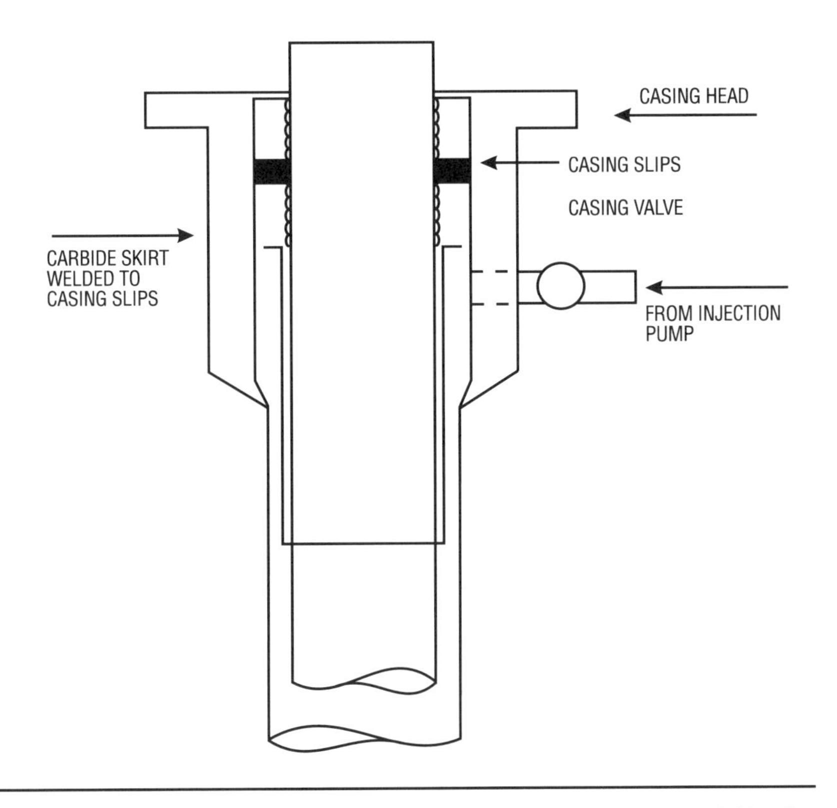

with a subsurface injection program may or may not be successful in the area of interest. Items such as area pressures and zone size can be reviewed from these offset disposal wells and determine what types of problems may occur during the proposed injection project.

One of the most important items that should be reviewed is the incidence of injection broaching to surface. The broaching of drilling wastes can occur at the wellhead or as far away as several hundred feet from the wellhead. If it is determined through this review that broaching has occurred in the field, then the reasons for the broaching should be investigated. Was the broaching caused by poor cement bonding, or was it caused by weak formation characteristics?

If through the review it is determined that the broaching was caused by poor cement bonding, then if the contractor feels comfortable that the cement bonding on the proposed well is adequate, he can probably rest assured that broaching will not occur at the site. If, however, it is determined that the formation in the area where the proposed well is located was the major reason for broaching in the offset wells, the contractor may want to reconsider the use of the proposed well as an injection well.

For offshore locations, where the formations are extremely young and unconsolidated, the chance of an injection zone broaching to surface is a lot higher than on land. The sure sign of an injection well broaching to surface in an offshore location is the discoloration of the water and the signs of bubbles somewhere around the rig.

When a well is on land — and since it cannot be determined the exact spot where the waste may broach to surface — the inadvertent broaching may cause substantial environmental damage to the area.

In a broaching well, once the fluid breaks down the formation, it will take the path of least resistance all the way to surface. This path of least resistance can be through a weak fracture plane at any distance from the well bore to commonly follow a path along the casing string. If abnormal geopressures are encountered, this could create a path for an underground blowout to occur. By conducting proper research of the field or area where the proposed injection is planned, this can be avoided.

When the contractor chooses to utilize a producing well for annular injection of drilling wastes, a great deal of research should be conducted to ensure that materials — such as the wellhead and casing(s) — between the injection zone and the producing zone are adequate to keep the two zones isolated. This becomes extremely important because damage to a producing well could cost a company millions of dollars in lost revenue and redrilling costs.

It is usually desirable to utilize a casing-to-casing annulus injection because of the lower injection pressures that will be realized during the injection operation. This is mainly because of the shallower zones that are encountered at the surface casing shoe and subsequently the lower fracture pressures at this point.

The biggest problem in deciding if a producing well can be used as an injection well is the shape of the casing. If a casing inspection log was run in the casing prior to completing the well for production, then unless corrosion has taken place in the casing string, the contractor can rest assured that the casing maintains the appropriate integrity to protect the producing zone or tubing-to-casing annulus from the injection pressures. If, however, no casing inspection log has been run, the contractor may have a difficult time in determining the casing integrity, because a casing inspection log cannot be run on a producing well. The contractor then will have to rely on the history of the well and the presence of any pressure on the annuli.

Anytime that a subsurface injection project is proposed for a producing well and the information regarding the mechanical integrity of the well is marginal at best, the contractor should weigh the chance of potentially damaging a money-making well against saving money on disposing the drilling wastes.

Before any decision on whether or not subsurface injection is the appropriate method for disposing of drilling wastes for a particular well, all of the alternatives detailing costs and risks should be explored. On land locations, a drilling contractor or operator has plenty of alternatives for disposing of drilling wastes. Hauling off, land farming, bioremediation, and injection are all viable alternatives. On offshore locations or in sensitive inland water locations, the number of alternatives is minimized: The drilling contractor is basically stuck with subsurface injection or hauling off, i.e., although for some offshore locations, overboard discharge of drill cuttings and drilling wastes still is allowed.

Other than the respective costs of the different alternatives, the contractor or operator should review the environmental impact each alternative will have. All alternatives, regardless of cost, should be focused on minimizing any damage to the environment. No matter how much money is saved by utilizing one technique over another, if the environment is adversely affected, a more environmentally safe option should be explored. This isn't just an altruistic approach. Paying for fines and cleanup costs related to environmental damage can far surpass any short-term cost savings.

If the alternative that is desired imposes a great amount of risk, such as the case of annular injection in a producing well, the contractor may opt to utilize another technique to dispose of the wastes. The cost of lost production or the cost of redrilling a well due to damage caused by a well bore failure can easily deem a field uneconomic to produce.

Prior to commencement of any subsurface injection operation, an injectivity test of the anticipated injection zone should be conducted to determine the capability of the zone to accept the fluid. Regardless of whether the subsurface injection will be down the casing through perforations or down the casing-to-casing annulus, an injectivity test should be conducted. Failure to determine the ability of the zone to accept the fluid could result in the injection operation to be aborted prematurely during injection or in causing damaging (to equipment or workers) equipment failures at the surface.

The surface equipment utilized to inject the waste in the well is a series of tanks, pumps and hoses. As with any equipment, the ability to conduct a certain operation with this equipment is limited by the pressure capacity of the equipment. Pressure spikes that can occur during an injection project due to solids bridging or zone plugging can cause catastrophic results during an injection project.

The injectivity test consists of hooking a pump to the casing valve or injection head of the well that is proposed for injection and water or drilling fluid pumped into the zone. The pumping should be done at different rates to determine the breakdown pressure and rate of the zone. The injection pressures at each rate should be recorded and a graph of the injectivity test plotted. Any abnormalities such as pressure spikes should be investigated to determine if they will pose a problem during the actual injection project.

Failure to determine the injectivity capabilities of the proposed zone could result in unexpected pressure failures of the injection equipment or the wellhead. In either case, the results could prove to be detrimental to the injection project or could cause serious harm to personnel performing the injection project at the surface. The cost in mechanical, environmental, or human terms of either one of these incidents occurring can be devastating. The mechanical failure of the surface equipment could result in the abortion of the project, thus costing the contractor thousand of dollars in equipment, slurrification, personnel, and additional disposal costs. The mechanical failure, such as a burst hose or cracked wellhead due to excessive pressure, could create a tremendous amount of environmental damage by the drilling waste. This could result in additional cost for remediating the site. A burst hose due to excessive injection pressure could injure the personnel working around the surface equipment during the injection project.

Only after all of the alternatives including the environmental aspects, risk and cost are explored, should an alternative be selected for the disposal of the drilling wastes.

SLURRIFICATION

Any time that a solid waste needs to be pumped from one location to another, it must be in a form that allows it mobility. Solid waste, regardless of its makeup, can be pumped once it goes through a process called slurrification. The process of subsurface injection requires that the solid drilling waste (drill cuttings, cement, et al.) be made into a slurry before it can be pumped or injected downhole. The restrictions at the wellhead, perforations, or surface pumping equipment and hoses dictate the size of particles that can be successfully pumped and injected into a well. In addition, if any waste is to be pumped, it must be in a liquid or semi-liquid form.

The slurrification process involves taking dry or semi-dry wastes and mixing them with a combination of water or another fluid and a suspending agent (Figure 9-3). In most cases the suspending agent used is bentonite gel. In the oil and gas industry, the process is called mechanical slurry disposal. When cuttings and larger pieces of debris are associated with the waste, and it is desirable to slurrify and inject the slurry subsurface, these pieces of debris first must be pulverized, ground, or crushed into a size small enough to be pumped through the surface equipment, the wellhead, or the perforations.

Since the advent of the slurrification process, several techniques have been developed to take the debris and convert them into a slurry. The equipment required to slurrify the mixture is relatively basic. The required equipment to mechanically slurrify wastes are a screen and hopper, a grinding unit, a shearing unit, storage tanks, centrifugal pumps, conveyor belts and a triplex injection pump. Although some contractors have utilized the rig equipment to mix the slurry, the wear and tear on the equipment due to the amount and type of wastes that are pumped can cost a tremendous amount of money in parts and labor.

The cuttings or debris are taken from their source and placed on the screen/hopper. The screen is usually a typical oil field shale shaker screen that can be changed for proper particle sizing. The hopper is typically a 5-10 bbl tub with sprayers to help minimize the amount of dust

FIGURE 9-3

SURRIFICATION SYSTEM

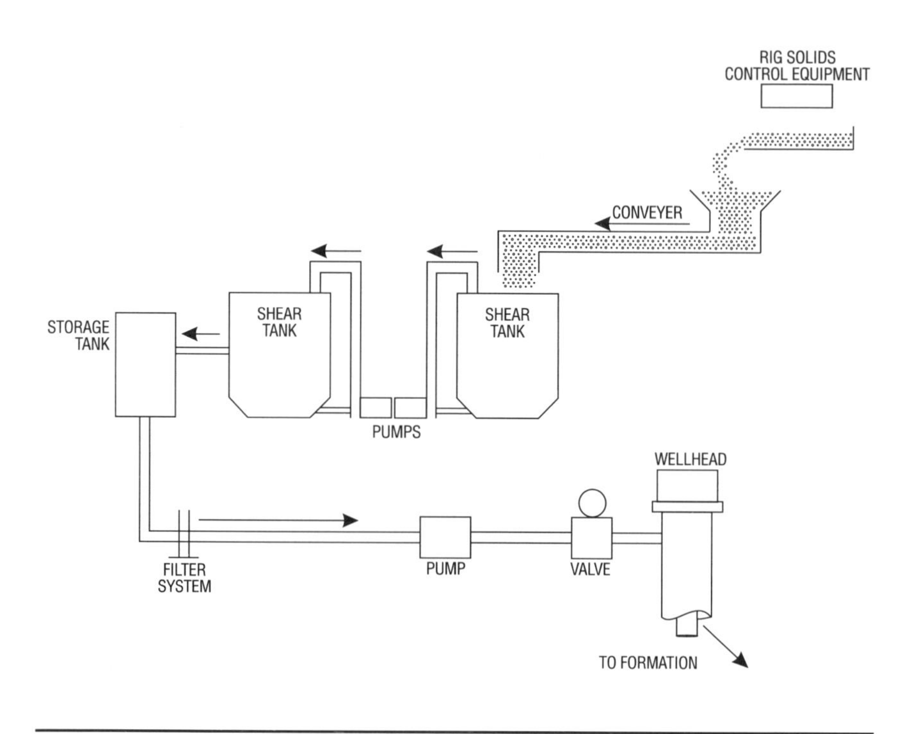

that is emitted into the atmosphere during debris loading.

The grinding unit is a specially designed two or three wheel carbide tipped grinder that is designed to grind the larger pieces of debris into pumpable sizes. The pieces of debris that do not pass the screening process are diverted into the grinder. The debris is then run across the screen. Any pieces that still do not pass the screen are sent back to the grinder for further pulverizing. This process is repeated until all of the debris will pass through the screen.

After the debris passes through the screen, it falls into a hopper and then onto a conveyor belt where it is dropped into a shearing unit. Water and the bentonite gel then are added to the dry debris and sheared into a homogeneous mixture. Water and gel are added to the mixture until at least an 80 section/qt viscosity is obtained. The mixture then is transferred between the two shearing tanks until proper slurry properties are obtained. The slurry is then pumped to a triplex pump where it is injected into the well bore or annulus.

The slurry mixture, after being processed through the mechanical slurrification unit, usually contains about 20 vol % solids. Through the screening, grinding and shearing process, the average particle size is about 100 microns. It is important that the 100 micron particle size be attained to prevent plugging of the injection zone. Unlike lost circulation material that contains a wide distribution of particles to build a wallcake, the particle size for injection purposes should be kept as homogenous as possible to prevent wallcake building and eventual formation plugging.

On a typical slurrification job, the contractor can expect at least a threefold increase in volume in the waste from the original dry waste volume. Therefore, when planning a subsurface injection operation, the receiving zone must be large enough to handle the slurrified capacity.

The slurrification process also is ideal when the contractor must dispose of oil-based cuttings. For onshore sites, the discharge of oil-based cuttings is strictly prohibited, so the contractor has only two alternatives when disposing of the cuttings. The contractor can choose to have the cuttings hauled off to an approved disposal site or slurrify the oil-based cuttings and inject them into a well. The first alternative can be quite expensive because the number of disposal sites that dispose of oil-based cuttings is dwindling as a result of stricter environmental regulations. The cost of disposing oil-based cuttings can be four times that of water-based cuttings disposal.

In the past, for offshore locations, the MMS allowed for oil-based cuttings to be directly discharged into the water. With increased scrutiny by offshore environmental regulations, discharge of oil-based cuttings and fluids into federal waters no longer is allowed. The slurrification process has given offshore contractors the ability to dispose of the oil-based cuttings in a manner that is less expensive than hauling off and is environmentally safe.

Critical safety and environmental concerns when using the slurrification process include increased vigilance around the grinding unit and around the wellhead and hoses due to elevated pressures. The grinding unit works much in the same manner as a paper shredder. Extreme caution should be taken to prevent personnel from accidentally getting their hands or clothing caught in the grinder wheels.

During the slurrification and injection process, the debris slurry is being pumped through a series of pipes, hoses, pumps, the wellhead and the casing. Although every step has been taken to reduce the size of the particles contained in the slurry, often pieces of debris of appreciable size remain in the slurry.

If the pieces are large enough or if they tend to accumulate, the possibility of them bridging off in the surface equipment, the hoses, the wellhead, or downhole is high. These bridges could cause pressure spikes during the slurrification and injection process that may exceed the pressure rating of the system's equipment. These pressure spikes, if large enough, could cause a failure in the equipment at surface or downhole. If a failure occurs at the surface, serious harm could occur to workers or equipment.

To help minimize chances of this occurring, pressure gauges should be mounted on the slurrification unit and at the wellhead and monitored throughout the slurrification and injection operation. Inspections and pre-operational pressure testing of the equipment should be conducted to ensure that all of the equipment will at the least meet the anticipated pressure during the mixing and injection process.

As was mentioned in the preceding section, an injectivity test of the proposed disposal zone should first be conducted to determine the amount of anticipated pressure during the disposal job. By knowing the injectivity characteristics of the proposed injection zone, a safe and effective injection project can be conducted.

Slurrification and injection recently has become an acceptable method to dispose of NORM. The NORM material is processed in the same matter that other drilling wastes are in using the mechanical slurrification system. Precautions to safeguard the personnel working on and around the unit are taken to ensure that inhalation or ingestion of the NORM material does not occur. Other than different regulatory requirements, the process for handling NORM remains the same as for non-NORM wastes.

HAUL-OFF

It may not seem like a modern technique in the handling of drilling wastes, but hauling off drilling wastes can be as effective as any of the scientific techniques used today. Miscalculating the type of waste(s) or the amount of wastes that are to be hauled off can prove to be costly, however.

With all of the rules and regulations that govern the hauling off of wastes today, the misclassification of a certain type of waste can prove to be costly to the contractor or operator. Although the oil and gas industry enjoys certain exemptions for handling drilling wastes, if through inadvertent actions or willful misconduct a waste that is considered as hazardous is mixed with non-hazardous oil field wastes, the whole shipment will be considered hazardous waste and must thus be treated as such, including its haul-off. Everything from how it is handled to how it is packaged for delivery to the disposal site will change once the waste is deemed hazardous.

If a waste is considered hazardous, the workers that handle the waste must be certified by Occupational, Safety and Health Administration (OSHA) training to do so and must be outfitted in safety gear before they can handle the waste. This is a big difference from the requirements of personnel that handle non-hazardous wastes. Typical rig personnel such as roustabouts and roughnecks do not have the appropriate OSHA training to handle hazardous waste, and thus the contractor would have to subcontract that service to an outside source. This could be costly.

If the waste is considered hazardous it must be properly containerized, tested, and labeled before it is hauled off. Instead of backing up a dump truck to the waste and loading it, it must now be carefully packaged in specially designed containers approved by EPA. The waste also must be laboratory tested to indicate the type and concentration of hazardous waste. Each package then must be properly labeled and sent to a disposal site that is certified and approved to accept and dispose of hazardous waste.

As a rule of thumb, it can cost 10-15 times more to dispose of hazardous waste than non-hazardous waste. Improper handling and disposal of the waste can prove to be a costly mistake. So it is critical that during any drilling operation that no mixing of wastes be done unless the two or more wastes to be mixed are known not to be hazardous wastes.

Closed-Loop Systems

A great deal of emphasis in this book has been placed on the handling and disposal of drilling wastes. With increasing government scrutiny and public awareness of environmental issues, the number of facilities that can receive wastes are being limited and techniques that once were thought of as state of the art and acceptable are now being banned.

This leaves the drilling industry with few affordable options to accomplish the goal of waste minimization and disposal. One such method that has gained popularity in recent years is that of closed-loop systems. The closed-loop system allows the drilling contractor to drill wells in environmentally sensitive areas without harming the environment as well as minimizing the amount of waste that is generated.

In the early stages of the environmental push in the oil and gas industry, closed-loop systems were mandated only in areas that were considered highly sensitive environmentally. These considerations were basically restricted to such areas as sensitive wetlands. Criteria such as the type of plants, the type of wildlife, and the proximity to special resources were all considered in the decision that no drilling wastes were to be discharged into the marsh and to mandate that a closed-loop system be employed.

Since this was a relatively new concept and few companies specializing in closed-loop systems existed, the first such systems were large, cumbersome, and costly. A drilling contractor could expect to add tens of thousands of dollars to the cost of drilling a well in wetlands because of additional equipment and labor required to install and operate the closed-loop system. In addition, since the intent of the closed-loop system was to prohibit the drill cuttings from being discharged into the marsh, the drilling contractor now had to assume the cost of disposing the drilling wastes generated.

As the use of closed-loop systems became more commonplace and the number of companies offering these services increased, the cost of installing and operating these systems were substantially reduced. Instead of piecemealing systems together from various parts of drilling

FIGURE 9-4

ONE-STEP CLOSED LOOP SYSTEM

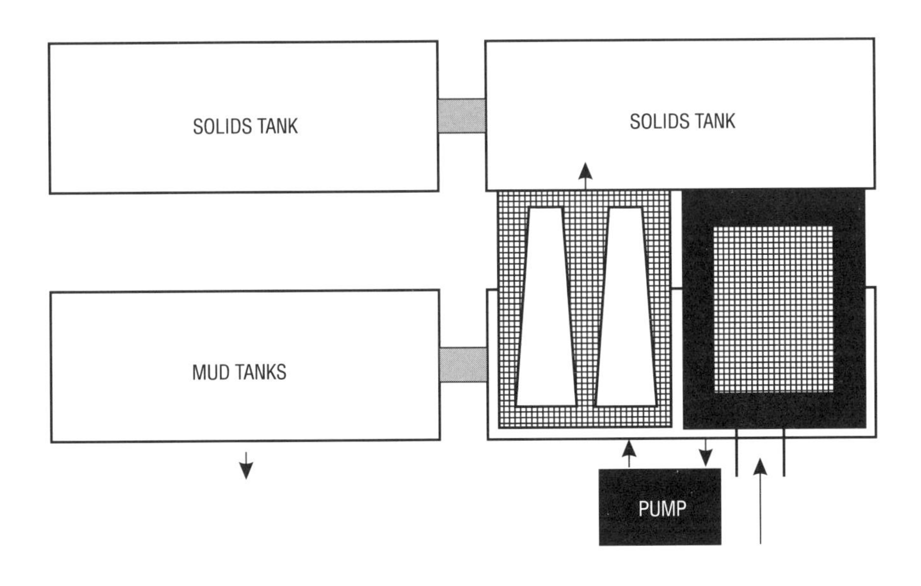

fluids/solids control systems, companies were now applying new technology and concepts to create modular systems that could be easily installed, operated and maintained. These new modular closed-loop systems also were more efficient. Many of the new closed-loop systems not only could minimize the amount of waste that had to be disposed of but also allowed the recycling of drilling fluid and make-up water for the drilling fluid. This added dimension would serve as a cost saving measure that would make the utilization of closed-loop systems more economic than not using the system.

The closed-loop system, when used for drilling wells, is designed to minimize the amount of waste that must be disposed of and eliminate the discharge of these wastes into the environment. The closed-loop drilling system consists of several individual parts of the existing drilling fluids/solids control system. A series of screened shale shakers, centrifuges, desanders, and desilters are utilized to remove as many of the solids from

FIGURE 9-5

MULTIPLE SHAKER/CENTRIFUGE SYSTEM

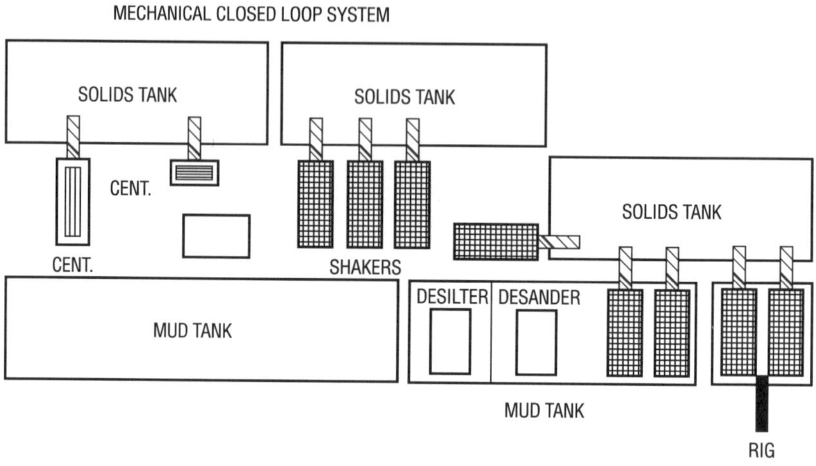

MECHANICAL CLOSED LOOP SYSTEM
SOLIDS TANK
SOLIDS TANK
CENT.
SOLIDS TANK
CENT.
SHAKERS
MUD TANK
DESILTER
DESANDER
MUD TANK
RIG

the drilling fluid so that the drilling fluid can be recirculated into the drilling fluid system for re-use. At the same time, the solids are being removed in a way that will help reduce the amount of liquids that are being discarded with them. The solids then are boxed or barged and sent to an approved landfill for disposal.

There are several versions of the closed-loop drilling systems on the market today. The earlier systems consisted of a series of shakers, centrifuges, and other solids control equipment designed to mechanically remove most of the solids from the drilling fluid. The more modern closed-loop systems employ a combination of the mechanical equipment and the chemical injection of the fluid to enhance the solids removal. In addition to the enhanced solids removal, the chemically enhanced closed-loop drilling system also was designed to recover much of the water that is used in the drilling fluid and use the recovered water as make-up water for the drilling fluid system. For wells in which the drilling fluids properties are not as critical, such as shallow, non-geopressured wells, a one-step solids controls system (Figure 9-4) was introduced. This one-step solids control system uses only a fraction of the equipment that the larger systems do and works as effectively on drilling fluid with a lower density.

The mechanical closed-loop drilling system is nothing more than an enhanced drilling rig solids control system. The system basically involves multiple shale shakers and centrifuges, strips many of the solids out of the drilling fluid, and returns the drilling fluid back into the system. The adage of "more-is-better" seems to be the philosophy in the design of this system. A typical solids control system on a drilling rig consists of one or more high speed shale shakers that are screened to remove the larger particle sized solids, a desander that contains two or more hydrocyclones designed to remove, the sand in the drilling fluids system that the high speed shale shakers did not remove, a desilter that contains six to twelve smaller hydrocyclones designed to remove the silt particles that the desander or shale shaker(s) did not remove and a centrifuge that is designed to remove the colloidal solids from the drilling fluids system that the other solids control equipment did not remove.

A centrifuge is not a standard part of a drilling rig's solids control system. Depending upon the type of mud system that is designed for the well or the depth of the well, the company that is contracting the drilling rig for the well may lease and install a centrifuge on the rig to help control the smaller particles in the drilling fluid.

One of the keys to successfully operating a drilling rig's solids control system is the contractor having the ability to dump and dilute the drilling fluid. After a certain portion of well bore drilled or time spent on a well bore, an appreciable amount of solids will build up in the system. In order to control an excessive amount of solids that build up, the contractor usually will dump a certain amount of the drilling fluids and build fresh drilling fluid. On offshore rigs drilling in federal waters, this is currently not a problem but on land and on inland water rigs (e. g. , wetland areas) this is not allowed or is becoming extremely restrictive. Therefore, utilizing typical solids control equipment to drill in a environmentally sensitive areas is becoming passe'.

By enhancing the drilling rig's standard solids control equipment with the addition of more shakers, larger de-sanders and de-silters, and one or two high speed centrifuges, the system can be converted into a mechanical closed-loop solids control system (Figure 9-5). The objective of a mechanical closed-loop solids control system is to dry the cuttings to as low a fluid content as possible while at the same time maintaining the percentage of drill solids in the drilling fluid as low as possible.

FIGURE 9-6

CHEMICAL FLOCCULATION SYSTEM

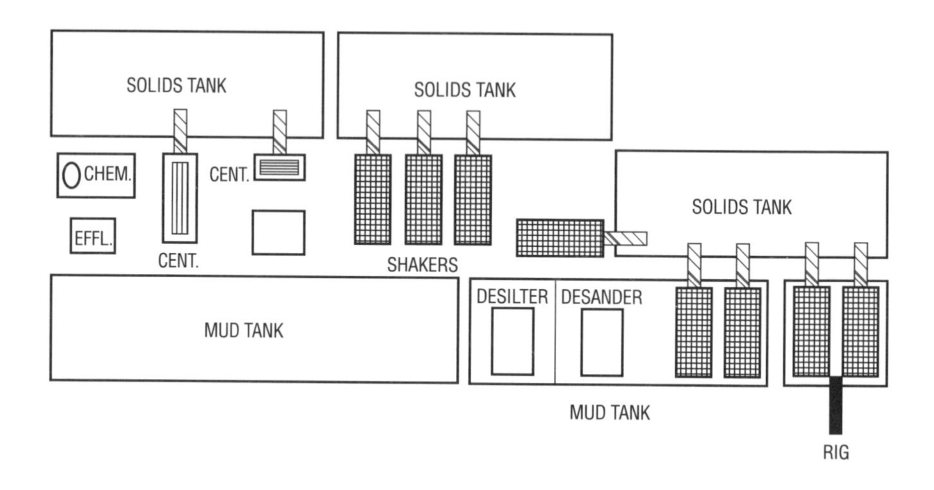

The drilling fluid that returns from the well bore contains the drilled cuttings. These cuttings range in size from microns to several inches in diameter. The return fluid goes directly to the first set of shale shakers where, through proper screening, the large pieces of cuttings are screened out and discharged into a cuttings box or tank.

The underflow from the first shale shaker is then run over another shaker that is screened to remove smaller particles from the drilling fluid. The cuttings are discarded into the same cuttings box or tank, and the underflow is deposited into a fluids tank.

The fluid then is picked up and run through the desander and desilter to help remove a larger percentage of sand and silt that the first two shakers did not remove. As before, the waste is discarded into the cuttings tank, and the underflow is deposited into another tank. A certain percentage of this fluid is then picked up by the centrifuge and processed to remove the colloidal solids that remain in the drilling fluid.

After the waste is discarded, the fluid is returned to the system and used to continue drilling the well. The problem is that the centrifuge only processes a small percentage of the drilling fluid. Although a substantial percentage of the solids in the system are removed by the mechanical closed-loop solids control system, the percentage remaining in the system can accumulate into a substantial amount.

As the drilling fluid is returned into the system for re-use, the remaining solids continue to be broken down into smaller particles that can begin to deteriorate the drilling fluid properties.

Furthermore, the more solids that remain in the drilling fluid, the more fresh water will need to be added to the system and thus the more chemical treatment will be required to maintain the desired drilling fluid properties. This basically means that the more fresh water that is required, the more fluid will need to be disposed of at the end of the drilling job. So although the mechanical closed-loop solids control system is effective, it does have its drawbacks.

If it is determined that the solids in the system are increasing and cannot be controlled by the mechanical closed-loop solids control system, the contractor or operator should consider the use of a chemically enhanced closed-loop solids control system (Figure 9-6). The chemically enhanced closed-loop solids control system utilizes much of the same equipment as the mechanical solids control system and adds a series of centrifuges and chemical flocculation to remove the smaller solid particles from the drilling fluid.

The chemically enhanced closed-loop solids control system removes the larger particles in the drilling fluid in the same manner that the mechanical closed-loop solids control systems does. The fluid runs over the shakers, through the desander and desilter, and then through the centrifuges. The effluent from the centrifuge then is put through another centrifuge, where a flocculating chemical is injected into it. With the combination of centrifugal forces and the chemical action of the flocculation chemical, the effluent is stripped of most of the colloidal solids that may remain. The effluent then can be returned into the system or stored in a storage tank and re-used as make-up water.

A big advantage that using the chemical flocculation system over any other system is that the effluent that is returned to the system or stored for make-up water for the building of more drilling fluid contains most of the chemicals that the drilling fluid has. This means a substantial cost savings in drilling fluids chemicals can be realized by utilizing this type of system.

The potential drawbacks of using a chemically enhanced closed-loop solids control system are the cost of the additional equipment that is required to operate the system and the cost of the flocculation chemical. Before deciding if this type of system is cost effective for the proposed operation, all of the costs associated with its application should be reviewed and compared with the alternatives. For instance, although the cost of the additional centrifuges and flocculation chemicals may be expensive, it should be weighed against the cost of having a treated fluid that already contains the drilling chemicals. When the contractor must build thousands of barrels of drilling fluid, the cost of treatment chemicals can be quite expensive.

Another intangible benefit of the chemically enhanced closed-loop solids control system is that the solids that are discarded into the cuttings tanks are extremely dry, thus minimizing the amount of waste that must be hauled off. This means lower disposal costs on the well.

When drilling deep wells where heavy drilling fluid densities are utilized, the cost of all of the special equipment such as the shakers, centrifuges and chemical injectors can be more than justified. But for wells that are relatively shallow (less than 12,000 ft) and where light drilling fluid weights are anticipated, (less than 12 lb/gal.), the use of a one-step closed-loop solids control system may be the solution. Unlike the more elaborate mechanical closed-loop solids control system and the chemically enhanced closed-loop solids control systems, the one-step closed-loop solids control system utilizes only a fraction of the equipment to accomplish the proper solids removal. With less equipment requirements, the lower the cost of the system is.

The basic concept of the one-step closed-loop solids control system is to process the total drilling fluids system rather than to process only a certain percentage of the system as is done in the larger systems. In the larger systems, only about 40-50% of the total drilling fluid that is returned from the well bore is processed through the entire system. Although the entire drilling fluids stream is run over the shale shakers in the larger systems, only a certain percentage of the fluid is processed through the desander, desilter, and centrifuges. This means that the solids remain in the system longer. Although the larger systems are designed to remove a large percentage of the solids through other means, the cost of these means can be expensive. By processing the total amount of fluid, the one-step closed-loop solids control system removes most of the solids initially, leaving very little solids, except for some colloidal solids, in the drilling fluids system.

The equipment for the one-step closed-loop solids control system are only a shale shaker, two 12 in. hydrocyclones, and a cuttings tank. It might be added that this is not in addition to the drilling rig's solids control system as in most of the larger closed-loop solids control systems but in place of the rig's equipment. The shale shaker, like on any system, removes a majority of the larger cuttings. The two 12 in. hydrocyclones pick up the total underflow from the shaker and through centrifugal forces removes a majority of the solids. The cuttings then are discarded into the cuttings tank and the processed drilling fluid returned into the system.

The concept of utilizing a large coned hydrocyclone for the removal of desired solids was first implemented in the mining industry. Because the particle size in the hydrocyclone can be carefully controlled through the adjustment of a concentric nozzle situated in the hydrocyclone, it was extremely desirable for the recovery of certain precious minerals. Not being different from any other type of hydrocyclone, fluid is injected into the hydrocyclones on the one-step closed-loop solids control system, and by centrifugal force, the solids and fluids are separated. The solids are discarded, and the fluids are sent to the system return line for re-use.

The key to the efficiency of the one-step closed-loop solids control system is that the hydrocyclones can process the total fluid in the system. A typical 12 in. hydrocyclone can process fluid at about 600 gal/min. When the hydrocyclones are run in parallel, a processing rate of 1,200 gal/min can be achieved. The typical pumping rate in the surface portion of the hole where the hole size is large and the hole is being drilled fast is 400-700 gal/min. On the large systems, only 200-300 gal/min of total fluid can be processed.

The use of the one-step solids control system also has proven to be effective in the removal of drill solids from the fluid without the addition of additional water into the system for dilution because of deteriorating fluid properties. This basically means that additional fluid disposal at the end of the job is minimized. In addition, the cuttings that are removed from the drilling fluid are extremely dry and can easily be hauled off to a disposal site or stored for later injection.

In addition to the benefits of efficient solids removal and waste minimization, the one-step closed-loop solids control system requires only a fraction of the space that the larger closed-loop solids control systems require. On average, the one-step system only requires one-fourth of the space that the typical larger systems require. With new environmental regulations limiting the amount of space that drilling contractors can utilize for drilling locations in environmentally sensitive areas, this system allows the drilling contractor to substantially minimize the size of the drilling location. With fewer oil companies installing large new drilling platforms offshore or attempting to utilize smaller existing platforms to drill from, the one-step closed-loop solids control system enables the contractor or operator to avoid making expensive modifications either to the rig or the platform to enable the installation of the closed-loop solids control system. In addition, the compact characteristics of the system allow for easy mobilization on a large crew boat or a small work boat.

There are some drawbacks to the one-step closed-loop solids control system, however. The most prevalent of these is that the effectiveness of the system is limited to lower density drilling fluids. Since the design of the hydrocyclones call for removing as many of the solids as possible, they tend to discard the barite that is in the drilling fluid. This means that through each processing cycle, barite has to be added to the system. This can be costly if the drilling fluid density is high and a large amount of drilling fluid is in the system. Since in lower density drilling fluids systems most of the fluid density is derived from the drill solids remaining in the system, barite is a minimal addition.

Another drawback to the one-step system is that because the 12 in. hydrocyclones are not designed to remove the colloidal solids in the drilling fluid, if well problems occur — thus increasing the time on the well or the amount of well bore drilled to more than was initially intended — the amount of colloidal solids tends to increase in the drilling fluid. If too high an amount of colloidal solids builds up in the drilling fluid, problems such as bore hole and fluid instability, pipe sticking, and reduced drilling penetration rates may occur. If either of these problems occur for even the shortest duration, the cost savings that may have been realized from utilizing the smaller, less expensive closed-loop solids control system can be quickly diminished.

If a contractor chooses to continue to utilize the one-step solids control system in these situations, this problem can be alleviated by the installation of a centrifuge in tandem with the dual hydrocyclones. The centrifuge, as with the larger systems, can help remove the colloidal solids that the hydrocyclones do not. During part of the operation, the contractor has the option of turning off the centrifuge and operating the system without it. All that is required to include the centrifuge into the circulating system, once it has been installed on the rig, is to turn it on.

The most modern one-step closed-loop solids control systems have been modularized where the tanks, shakers, hydrocyclones and the back hoe are all on one skid. This enables easy mobilization to land, inland water and offshore locations.

Waste Handling and Minimization Overview

After reviewing the numerous options for waste disposal that are available to the contractor, it is apparent that conducting proper research as to the type of wastes, quantity of wastes, and objectives is extremely important.

Each site will have different needs to contend with and thus a combination of the disposal options to effectively and efficiently dispose of the wastes may be necessary. On land, unless a complete closed-loop drilling fluids system is utilized, at least one earthen reserve pit will be utilized. In the shallow portion of the wells where fluid properties meet all regulatory criteria and the land owner allows it, all of the drill cuttings and drill fluid can be land farmed as a form of disposal. This saves the contractor from having to either haul off or inject this waste subsurface. For most wells, this is about 70% of the waste that is generated by drilling the well. As for drilling wastes that are generated in the deeper portions of the well, the wastes must be hauled off or injected. For most wells this includes all of the drilling fluid, the drill cuttings and emergency pit wastes.

On any particular drilling location, a variety of drilling wastes will be generated. This being the case, a variety of waste disposal methods will need to be employed at any one location. In addition, several different government regulatory agencies will share in the responsibility of enforcing the environmental regulations.

The key to being successful in efficiently and economically minimizing, handling and disposing of drilling wastes is properly characterizing, qualifying and quantifying the wastes. With this knowledge, the contractor can determine the type of technique that will best fit the handling and disposal objectives. This methodical approach to handling and disposing the drilling wastes can ultimately save the contractor hundreds of thousands of dollars on each drilling job.

On a typical offshore location, unless the contractor is utilizing oil-based drilling fluid, disposal of the drilling fluid, drill cuttings, waste water and sanitary waste can be discharged into the water. The domestic wastes such as garbage are usually compacted and sent onshore for disposal.

On inland water locations, since no wastes can be discharged into the surrounding area, all of the wastes are run through a closed-loop solids control system or placed on and in barges and taken to an approved disposal site.

Table 9-1 is an example of the type of drilling wastes that can be expected when drilling a typical twelve thousand foot deep well in South Louisiana, and the common techniques that are utilized to minimize, handle and dispose of the wastes.

TABLE 9-1

Typical Drilling Wastes For A 12,000 Ft. Well In South Louisiana

Drilling Depth, ft	Waste type	Amount of waste, bbl	Handling disposal method
0-3,000	Drilling fluid	900	Maintain in system. excess fluid, reserve pit and land farm
	Drill cuttings	900	Reserve pit and land farm
	Waste water	400	Treating pit, discharge to land farm area
	Cement	50	Emergency pit, retard with sugar and haul off or maintain for injection
	Garbage	(0. 5 tons)	Dumpster and hauled to landfill
	Sanitary waste	100	Treat in system and discharge into ring levee, haul off or inject
3,000-9,000	Drilling fluid	900	Maintain in system, excess fluid: reserve pit and land farm
	Drill cuttings	900	Reserve pit and land farm
	Waste water	1000	Treating pit and discharge to land farm area
	Garbage	(1 ton)	Dumpster and haul to landfill
	Sanitary waste	200	Treat and discharge into ring levee, haul off or inject.
9,000-12,000	Drilling fluid	250	Maintain in system. Excess fluid into reserve pit. Store for re-use, injection or haul-off.
	Drill cuttings	250	Reserve pit and store for haul-off, slurrification and injection or trenching and burying.
	Waste water	1000	Treating pit and discharge to land farm area.
	Sanitary waste	400	Treat and discharge into ring levee, haul off or inject.
	Garbage	(1 ton)	Dumpster and haul off to landfill.
	Waste oil	10	Place in drums and send to recycling center.
Completion	Drilling fluid	450	If not packer fluid, re-use on other site, inject or haul off.
	Brine	600	Maintain in system. Inject or haul off to disposal site.
	Produced sand	(0. 5 tons)	In tanks, NORM check. Inject or package and send to disposal.
	Waste oil	5	Place in drums and send to disposal site or recycling center.
Post- well operations	Location boards		Re-use, landfill or forced air burned.
	Creosote boards		Re-use or special landfill disposal.
	Pits		Closed. Mixed to local, state, or federal requirements. If contaminated, soil treated or sent to disposal site. Back filled.
	Land farm site		Tested. Tilled and reseeded.
	Excess fluid		Hauled off or injected.

REMEDIATION

Because almost all industrial processes develop byproducts in the form of waste that cannot otherwise be handled or disposed of economically, remediation becomes inevitable. The drilling industry is no exception. When reviewing the drilling industry, we tend to focus mainly on the drilling fluid and the chemicals that are used in it. In reality though, there are several functions in the drilling industry that also have environmental effects. In addition to the drilling fluid, there is the effect of oil leaks from the mechanical parts of the rig, the reserve pits, washdown water from the rig floor, and the drilling location itself.

We often hear that the solution to pollution is dilution. In fact it is but one of many ways of remediating certain types of contamination. Dilution should never be used when the contaminant has been classified as a hazardous waste. As the law reads, if you add as little as a teaspoon of hazardous waste to a mountain of non-hazardous waste, the whole lot will be considered as hazardous waste. But when enough non-contaminated soil or water is available for mixing to reduce the contaminant to acceptable levels, then dilution is an economical way to remediate a site.

Most drilling reserve pits can be remediated by the use of the dilution technique. Before a contractor or operator decides to resort to dilution to remediate a pit, he must first determine the contents of the pit. As was mentioned, if any hazardous waste was inadvertently placed in the pit, the whole pit may be deemed hazardous and special techniques to haul off the contents and remediate it will be necessary.

If a contractor or operator opts for dilution to remediate a pit, he must first ensure that enough uncontaminated soil — preferably adjacent to the pit — is available to successfully mix with the contents of the pit in order to bring the composite levels within government specifications.

Determining the amount of soil required for dilution is a fairly straightforward process. All that is needed is the amount and concentration of contaminated soil and the amount and concentration of non-contaminated soil. These concentrations can be determined by obtaining a soil sample and testing it for the desired parameters or contaminants.

Once the soil compositions and quantities are known, the dilution ratios can be calculated. The calculation for diluting non-hazardous drilling wastes is a straightforward algebraic equation. For instance, if there is a pit that contains 100 bbl of barite contaminated soil with a concentration of 100,000 ppm, it will take 233 bbl of non-contaminated soil mixed with the contaminated soil to bring the concentration down to 30,000 ppm. In equation form this is:

Vi =	100 bbl; Ci = 100,000 ppm
Vf = ?	Cf = 30,000 ppm
and since	Ci x Vi=Cf x Vf
Vf = (Ci x Vi) / Cf	
then	Vf = (100,000 x 100) / 30,000

Vf = 333 bbl of total mixture at 30,000 ppm

or 233 bbl of non-contaminated soil must be added to the contaminated soil to bring the concentration to the desired level.

From this calculation, the contractor can calculate the amount of acreage required for a proper dilution. The total amount of acreage will first depend on the availability of land; second, on the concentrations of contaminants in the soil as determined by the background sampling; and finally, by the depth the contractor intends to till the mixture together. The same method can be used on any non-hazardous contaminant assuming that the volumes are additive.

If after conducting the pre-remediation evaluation of the site and waste, not enough dilution area is available, then the contractor may have to consider a partial haul-off or total haul-off of the wastes. For instance, if there is only 1 ac of uncontaminated land available for dilution or mixing and there are 5,000 bbl of 160,000 ppm barite contaminated soil, it probably will be necessary to haul off 4,700 bbl of the contaminated soil and thus be able to land farm only 300 bbl of it.

In most drilling reserve pits there will be "hot spots" of certain chemicals that contain extremely high concentrations. In a typical shale pit, a hot spot of barite or chromium is usually common. This is prevalent mainly because the discharge trough from the drilling fluids system is situated in one position and usually remains there throughout the drilling operation. At times the need to discard a certain volume of drilling fluid because of over-inventory will result in abnormally high concentrations of a certain chemical. This is typically the weighting material that is used to increase the density of the drilling fluid. Fortunately, unless oil-based drilling fluids are being used on the rig, barite is not considered a hazardous waste and can be diluted to achieve the proper soil concentrations. If other chemicals are being utilized that are not exempt oil field wastes, the contractor or operator may have to utilize hazardous waste remediation techniques to clean up the waste.

A contractor should plan, for any drilling reserve pit remediation, to haul off a certain portion of the soil in the pit in order to make the dilution process easier and more effective.

After mixing the waste with the soil, it is critical to conduct post-remediation sampling to ensure that proper mixing has been accomplished. If the post-remediation analysis indicates that the soil mixture is still high in a certain concentration, the contractor or operator may have to haul in more non-contaminated soil onto the location and conduct more mixing in order to bring the concentrations to acceptable levels.

If the company cannot bring the concentrations to state or federal levels or to background levels, whichever are higher, it may be required to scrape up all of the soil and haul it off to a disposal site as well as bring in fresh, non-contaminated soil to replace the removed soil. This is a perfect example of why it is extremely important to conduct a thorough pre-remediation analysis of the site and thus save the company as much as tens of thousands of dollars in disposal costs.

The proper functioning of a drilling rig relies heavily on its lubrication. It has hundreds of moving parts that must be lubricated properly, or the result is expensive mechanical breakdowns of the equipment and costly down time to repair or replace these parts. But this also means that there will inevitably be lubricating material (oil and grease) spillage on the site. The amount of spillage can be minimized by installing drip pans under all of the equipment and using lubricating material only where it is required. The amount of spillage also can be minimized by use of a diesel-electric drilling rig rather than a mechanical drilling rig. The diesel-electric drilling rig relies on generators to provide most of the drilling rig's power compared with a mechanical drilling rig that relies on diesel engines, belts, and chains to convey the hoisting and rotating power to the drilling rig's components.

By using a mechanical drilling rig, hundreds of gallons of oil and grease can be inadvertently spilled on the drilling location. Most of the time, however, the total impact of the spillage is not known until drilling has been completed and the rig has moved off location. By this time it is too late to employ preventative measures to minimize spillage.

One way that a company can ensure that the impact of spilled oil and grease is minimized is to place an impermeable liner below the location boards. A 20-40 mil liner placed under the location can reduce the chances of contaminating the soil with oil and grease. This is not to say that sound environmental practices can be ignored, but the placement of the liner can assist in this area.

After the well is drilled and the rig is moved off location and/or if the location was boarded, and the boards removed, the company now is faced with cleaning up the oil and grease contaminated soil. In many cases, much like with barite contamination, if enough land is available for mixing, the oil and grease can be worked into the soil and the concentration brought down to acceptable levels (1% oil and grease is usually an acceptable level).

But in the case where there is not enough non-contaminated land available for this operation, non-contaminated soil may need to be trucked in for mixing, or another alternative to remediate the oil and grease contaminated soil may be required.

There are basically four alternatives available if soil mixing or land farming of contaminated soil is not feasible: slurrification and injection into a well bore, haul-off to an approved disposal site, bioremediation, and chemical remediation.

Slurrification and injection into a well bore allows the drilling contractor to utilize the current or offsite well bore to safely, easily, and inexpensively dispose of the oil and grease wastes. If the oil and grease wastes are not considered hazardous wastes or if they have not been mixed with hazardous wastes, the oil and grease can be mixed with the drill cuttings, drilling fluids, and other drilling wastes that are earmarked for disposal via slurrification and injection.

Haul-off of the oil and grease contaminated soil can become quite expensive depending on two factors: trucking and disposal cost. Depending upon the location of the well, trucking costs could quadruple the cost of disposal and if interstate trucking is involved, maybe even cost more. One thing to keep in mind when dealing with oil and grease waste, however, is to consider depending on the amount of oil and grease wastes, having the oil and grease waste tested for its BTU (British Thermal Unit) content. If the BTU content of some wastes is above a certain level, some disposal sites will pay the generator, or at least charge only a small fee, to haul off the waste. This is because waste with a high BTU content can be utilized as fuel to operate disposal site incinerators. So it may be worthwhile to check the BTU content of the oil and grease waste before another alternative is utilized to dispose of or remediate the waste.

If a contractor has enough time, bioremediation can be used. Bioremediation techniques utilize cultured bacteria to assist in the natural processes that break down hydrocarbons. Unfortunately, most bioremediation treatments require 6-8 months to work properly and require a fair amount of maintenance. In most cases the drilling contractor or operator does not have this kind of time to allow the bioremediation product to work because the land owner is ready to begin using the land almost immediately after the rig has been removed. Only if the location is situated in an area that does not have a time restriction should bioremediation be utilized.

The fourth alternative is to employ chemical remediation. Chemical remediation combines the efficiency of hauling the waste off to a disposal site or subsurface injection with the effectiveness of bioremediation. The process involves using a sodium silicate based chemical on the contaminated soil. The sodium silicate is designed to break the bonding between the hydrocarbon and the soil, thus releasing the volatile hydrocarbons. As with bioremediation, some maintenance is required, but as with haul-off, the process only requires a few days to administer.

In order for the chemical remediation technique to be effective, the soil on which it is being applied must be broken down into small particles. This can be accomplished either through a process called soil washing or a process using a hammer mill.

The soil washing technique basically utilizes rig-styled equipment, including a shale shaker, desanders and desilters, and centrifuges to break down the contaminated soil. Once the contaminated soil is broken down, the soil then is treated with the chemical and piled or spread out over a designated area. One problem with using a soil washing unit is that after the soil has been washed, there will be a large amount of effluent to be disposed of. If the drilling contractor has a discharge permit to allow the discharge of the effluent from the soil washing unit, this is no problem. But if the drilling contractor does not have a discharge permit or the site is located in an area that does not allow the discharge of any fluid, such as a sensitive wetlands area, the drilling contractor may be stuck with having to haul off the effluent to an expensive disposal site.

The hammer mill technique utilizes a large grinder to break down the contaminated soil and help release the volatile hydrocarbons contained in the soil. The broken down soil then is put on a conveyor belt and spray treated with the sodium silicate chemical. The treated chemical then is piled on the location and spread out to dry. The advantage that the hammer mill process has over the soil washing process is that no water is added to the soil and thus no additional effluent is created. This means that the drilling contractor does not have the problem of having to dispose of any fluids after the hammer mill treatment process.

On a typical 400 cu yd (about 2,000 bbl) hydrocarbon- contaminated soil site, the contaminated soil can be treated and readministered in about 10-14 days. This process has been successful in reducing oil, grease, and other hydrocarbon contamination levels from 20,000 ppm to below detectable limits for about one third the cost of other remediation techniques. Furthermore, unlike bioremediation techniques, this process requires only one tenth the time of other remediation techniques. This keeps the land owners happy and gets the contractor on to another location, making money.

Another common type of contamination that is common on a drilling site is that of saline or brine water contamination. In areas where fresh water is scarce, brine water often is used as make-up water for the drilling fluid. The salinity of the water can be 500-200,000 ppm, depending upon its source.

Although special precautions are taken to store the make-up water in open top rig tanks or enclosed frac tanks, few precautions are taken with the drilling fluid in which the water is used. The earthen pits are not typically lined, and thus post-drilling contamination of the pits is common. The effects of salt contamination can be as devastating as any other form of contamination. Unlike hydrocarbon contamination, unless the contamination is located in an area of repetitive tidal fluctuations, natural bioremediation will not eventually remediate the salt damage. Very few types of vegetation can be grown on land that is salt-contaminated.

There are currently three methods of remediating salt-contaminated soil: haul-off, subsurface injection, or chemical treatment. Land farming is usually not effective or recommended in remediating salt contamination.

As with other types of contamination, haul-off can be quite expensive. On a typical 300 ft by 300 ft by 7 ft pit, haul-off of salt-contaminated soil can cost — including dredging, trucking, and disposal — as much as $1. 5 million. This is the cost of drilling most 12,000 ft wells. So unless other alternatives are not available, one of the other remediation methods should be reviewed.

Subsurface injection, as was discussed earlier, also is a popular method. As with oil and grease or barite contamination, this depends much on whether or not there is a well that can be utilized for this purpose.

Chemical treatment offers the advantages of haul-off and subsurface injection. Salt contamination can be chemically treated with a calcium nitrate chemical. The calcium nitrate releases the sodium bonding in the soil and thus reduces the salinity of the soil to an acceptable level. Unlike haul-off, utilizing this chemical can be economical and efficient. And unlike an injection well, you are not dependent upon the mechanical integrity of well equipment.

There are two methods of applying the chemical treatment method to a salt-contaminated soil. The first method calls for the contaminated soil to be excavated. After the soil is excavated, a treatment of the calcium nitrate chemical is made on the bottom of the excavation. A 1. 5-2 ft layer of soil then is placed in the excavation. The freshly placed soil then is treated with the calcium nitrate chemical. This layering and treating of the soil is repeated until all of the soil has been replaced in the excavation. This will allow the chemical to make direct contact with the contaminated soil and begin working immediately. This is the most efficient way of remediating a salt-contaminated soil with this chemical. If applied properly, the salt-contaminated soil will be remediated within 6 weeks at only a fraction of the cost of hauling off to a disposal site.

If a company has more time for remediating the salt- contaminated site, the calcium nitrate chemical can be applied by drilling several holes throughout the property, constructing a berm around the area and flooding the property with the chemical. This technique is designed to allow the chemical to soak or seep into the contaminated soil over time.

The problem with this technique is that, depending upon the depth of contamination, it could take several months to remediate a site. In addition, the effectiveness of this technique relies heavily upon the permeability of the soil. If the permeability of the soil is low, and the soil is not homogeneous, several areas may not ever come in contact with the chemical. This may lead to fingering of the treatment and could lead to dead spots on the surface.

Other than soil or water contamination, if a well site is drilled in an area where a board location is necessary, then the problem with the disposal of the boards can be a major one. On a typical 300 ft by 300 ft turnaround, taking up the boards and properly disposing of them can be a major undertaking. Once again we are faced with either hauling off the boards or attempting to dispose of them onsite. Often, a contractor will attempt to bury boards. This works up to a point, but when there is more than 90,000 sq ft of boards this can be difficult.

Haul-off of boards, because of trucking and disposal costs, can be quite expensive. There is one technique, however, that can be used that is economic and effective. Forced-air burning onsite. Forced-air burning is a technique that is accepted by most government agencies. This is a popular method with environmental agencies because it is designed to minimize any air emissions. It is popular with contractors because it minimizes costs and does not require any special permit to conduct it.

This method entails digging a 6-10 ft pit, placing the incinerator unit over the pit, placing a curtain around the unit, and feeding the boards into the unit. The unit relies on forcing air into a flame and thus burning at a temperature that minimizes the amount of smoke that is released from the unit. All that is left to dispose of after the boards have been burned is a few pounds of ash and a few nails.

If none of these techniques appeals to the company that is utilizing boards on a location, then it should consider spending a little bit more money up front and utilizing mats for the location boards. The interlocking mats are usually 3-4 ply thick and can withstand most of the heavy loads imposed on them by drilling rig equipment and trucks. This enables the contractor to just take up the boards at the end of a well and utilize them on the next site.

DRILLING AND THE FUTURE

BALANCING ACT

The U.S. drilling industry's future health depends to a great degree on how long it can continue to accommodate growing environmental concerns without committing economic suicide in the process.

The balancing act, in turn, depends on whether government — especially at the federal level — continues to apply some degree of reason to enforcement of environmental regulations.

The U.S. drilling industry has dodged the bullet on having most E&P wastes come under the purview of RCRA Subtitle C for a number of years now, but that does not mean it is home free. What may be overlooked by most casual observers in looking at how EPA classifies exempt vs. nonexempt substances is that the EPA's exemption under Subtitle D of many E&P wastes from its hazardous list is only temporary.

At the same time, there is an ambiguous quality in how EPA classifies certain wastes, using qualifiers such as "possible" or "believed to be." So an operator or drilling contractor cannot truly be certain that a substance believed to be exempt today might not at some point in the future fall under the onerous regulatory regime of Subtitle C, SARA, Cercla, and the other laws that come into play regarding hazardous substances.

Meanwhile, the clamor including E&P wastes under the nonexempt listing has not completely died down, and a number of environmental groups and some state environmental agencies have made the inclusion a priority.

In its 1988 finding that most E&P wastes were exempt from Subtitle C consideration, EPA also found that while "existing state and federal regulations are generally adequate. . . Certain regulatory gaps do exist, and enforcement of existing regulation in some states is inadequate. "

To deal with those concerns, EPA disclosed a three-pronged approach that calls for:

- Improving federal programs under existing statutory authorities in RCRA Subtitle D, the Clean Water Act, and the Safe Drinking Water Act.
- Working with states to encourage improvement in the states' regulations and enforcement of existing programs.
- Working with Congress to develop any additional statutory authority that may be required.

In seeking to help EPA with the second prong of the agency's approach toward E&P wastes, the Interstate Oil Compact Commission (now the Interstate Oil and Gas Compact Commission) formed the Council on Regulatory Needs in January 1989. The council in December 1990 put out a report in which it recommended effective regulations, guidelines, and/or standards for state level management of oil and gas drilling and production wastes, with EPA's concurrence.

The IOCC Council on Regulatory Needs published these recommendations, along with a survey of member states' regulatory guidelines covering E&P wastes.

While the council's report was a commendable effort to bridge the regulatory gaps that EPA found at the state level, it also was criticized at the time as not going far enough to regulate E&P wastes.

A minority report to the Council on Regulatory Needs concluded that council's report to IOCC:

". . . is essentially a restatement of the status quo, a reaffirmation by the states of practices they already allow, regardless of whether those practices are protective of public health and the environment.

The Council did not move the participants to write a report that suggested the best possible regulatory programs — a set of guidelines that would move both the states and the industry forward.

Instead, the report is further justification for continued minimal action. In conjunction with our recent rereading of the damage cases in

(EPA's regulatory determination on E&P wastes), the council's effort simply convinces us even more of the need for a stringent federal regulatory program for E&P wastes.[1]"

The minority report goes on to detail what it sees as regulatory inadequacies in EPA's and IOCC's approach to dealing with oil and gas E&P wastes, but the point to be made here is this: the drilling industry may have won the key battles to date on the issue of exempting most E&P wastes from hazardous status, but the war is far from over.

Industry's Challenge

With the likelihood that all of the elephant fields in the United States have been found, the domestic drilling companies face a tremendous challenge. Drilling in water depths of 5,000 ft could become commonplace, but if low oil and gas prices linger, that will not be economic. The same goes for companies having to drill deeper wells on land. The specialized equipment required for the harsh environment of a 25,000 ft well is neither readily available nor inexpensive.

But these are just a couple of the challenges that the domestic oil and gas industry faces in looking towards the 21st Century. The biggest challenge appears to be the increased scrutiny that is being imposed on the industry through new environmental regulations. These regulations not only govern how a well is to be drilled but where.

This increased scrutiny costs the industry hundreds of millions of dollars each year in operating costs as well as limiting the manner in which drilling can be accomplished. This increased cost is making profitable projects marginal and already marginal prospects non-profitable ventures.

This is why any technological challenge facing the drilling industry is dwarfed by the challenges the industry faces in environmental and regulatory compliance.

As oil and gas industry professionals, we all have the obligation to meet these challenges. We have already accepted the challenge of providing the nation with its energy needs and now must determine how to, through creative approaches to drilling, explore for oil and gas with the understanding that the regulations that govern our operations are extremely restrictive. Failure to meet these challenges as an industry can have devastating consequences to the energy and economic security of the U.S.

Along with meeting the technological challenges, we must take it upon ourselves to become involved in the legislative process that is creating the rules and regulations the oil and gas industry must live by. The adage of "The squeaky wheel gets the grease" holds especially true for the oil and gas industry. By not lobbying to have laws passed that will benefit our industry, we will have to live with laws that are mainly influenced by non-industry concerns. Although only a few states have a substantial amount of oil and gas production, the economic well-being of each state depends to some degree on health of the domestic oil and gas industry.

Regulations have been proposed that could eliminate many of the much needed regulatory and fiscal considerations the oil and gas industry currently enjoys. Many of these proposed changes stem from a lack of understanding by the general public of the operating practices of the industry, negative publicity from isolated cases, and the importance of not becoming dependent on foreign energy.

As an industry, efforts must be made to educate the general public about the operating practices required in drilling a well. The public must be educated about the minimum space needs of a drilling location, the types of wastes that are generated at these sites, and the impacts these operations will have on the environment. Having a well-informed public is the best defense the oil and gas industry can have.

While the First Amendment to the U.S. Constitution guarantees freedom of speech, that right is at times abused by the media. Unfortunately, a few isolated incidents involving accidental oil spills determine how the U.S. oil and gas industry is viewed by the general public. It is not just that the accidents occurred as reported, but the media does not always tell the whole story — often to the industry's detriment. We cannot manipulate how the news is reported, but we can best avoid more negative publicity by seeing to it that we minimize the opportunity for these accidents to occur by implementing sound environmental practices in all of our operations. Although not all accidents can be prevented, they definitely can be reduced.

The U.S. oil and gas industry must cope with trying to meet the ever increasing energy demands of our nation. With the business arena becoming more global, it is vital that those involved in global business have the ability to do so and can afford the cost to compete in this environment.

Becoming more dependent on foreign energy is a threat to the security of our nation, not to mention an expensive alternative. As we found with the Arab oil embargo of the early 1970s, such dependence on others for our supply of oil can be crippling to the mobility of our nation.

It is apparent that the battles between the oil and gas industry and environmentalist groups are far from over. It is extremely important to understand that we all have an obligation to safeguard the environment from damage, but it is equally important for the domestic oil and gas industry to have the ability to supply the nation with energy for the future that is both plentiful and affordable.

1. EPA/IOCC Study of State Regulation of Oil and Gas Exploration and Production Waste: A Project of the Interstate Oil Compact Commission's Council on Regulatory Needs, December 1990.

INDEX

S

Z